Physics and its Laws

Advanced Level of Physics

Published by:
    Alfie Law

Contents

An example of the laws of physics: The law of gravity 3

The relationship between mathematics and physics 40

Important Conservation Principles 81

In the laws of physicssymmetry 120

Distinguish between past and future 172

    Figure 26 189

    Figure 27 190

Probability and uncertainty nature related to quantum mechanics 205

    Figure 28 212

    N12 = N1 + N2 (no interference) 217

    h12 = h1 + h2 but 223

    Figure 30 224

    Figure 31 229

The search for new laws 245

    Figure 32 247

    Figure 33 252

An example of the laws of physics: The law of gravity

I'll start with something that has always struck me as odd. On the rare occasions when I'm asked to play bongos at a large gathering, the moderator doesn't need to mention that I'm a theoretical physicist. I think it's because there's more respect for art than for science. Renaissance artists believed that man's primary preoccupation was man.

There are other interesting things in the world too, although they say it should be. Even artists notice the beauty of sunsets, sea waves and stars in the sky! Observing these is enough to give us an aesthetic pleasure. There is a rhythm and order among natural phenomena that is invisible to the naked eye but can be noticed when viewed with an analytical eye. What we call the laws of physics is this rhythm and order itself. In this lecture series I would like to go into the general properties of the physical laws. This is another level of generality; a generality in which we can say that the laws are superior to them. The subject I will deal with will be nature, which we see as the result of a detailed analysis; However, I will speak only of the broadest general properties of this kind.

A subject containing such generalizations tends to turn to philosophy; Conversations can be experienced as "deep philosophizing". I will take a more specific approach, preferring to be truly understood rather than vaguely. In this first conference I will drop generalizations and talk about a specific law of physics. Therefore, I will give an example of what I will discuss later in general. I will use this example as needed to make something that might come across as very abstract more concrete. As a concrete example of physical laws, I have chosen the law of gravitation, the phenomenon of gravitation. I don't know why I made this choice. This was one of the first fundamental laws discovered and has an interesting history. Now you can tell me, "That's an old story; I would like to learn more about modern science." Perhaps we can speak of "newer"; but one cannot speak of 'more modern'. Modern science follows the same tradition as the discovery of the law of gravitation. Therefore, we only talk about recent discoveries. I don't think the law of gravity is a bad choice; because I am treading a very modern path in describing their history, their methods and the nature and quality of their discovery. you can't speak more modern'. Modern science follows the same tradition as the discovery of the law of gravitation. Therefore, we only talk about recent discoveries. I don't think the law of gravity is a bad choice; because I am treading a very modern path in describing their history, their methods and the nature and quality of their discovery. you can't speak more modern'. Modern science follows the same tradition as the discovery of the law of gravitation. Therefore, we only talk about recent discoveries. I don't think the law of gravity is a bad choice; because I am treading a very modern path in describing their history, their methods and the nature and quality of their discovery.

This law is considered "the most sweeping generalization ever made by human intelligence." However, you can guess from my previous words that I am more interested in a natural wonder realized by an elegant and simple law like gravity than in human intelligence. So we don't focus on that we are smart enough to spot it, but that nature is smart enough to take it into account.

The law of gravitation states that two masses exert a force on one another that is inversely proportional to the square of their distance and directly proportional to the product of their masses. We can express this important law with the following mathematical formula:

So it's a constant number times the product of the masses over the square of the distance.

$$F = G\frac{m \cdot m'}{r^2}$$

Now if I add that a mass is accelerated by the action of a force, or that its velocity changes every second in inverse proportion to its mass, or that as mass decreases its velocity changes more in inverse proportion to mass, then all are Law of Gravity requirements met. Anything beyond that is just the mathematical result of those two things. I know that you are not all mathematicians and that you cannot immediately see all the mathematical consequences of these two things. For this reason I am going to give you brief information about the history of the discovery, some of its consequences, its impact on the history of science,

Briefly, the story is as follows: Ancient scholars observed the movements of the planets in the sky and concluded that they revolved around the sun with the earth. This result was also later discovered independently by Copernicus - people had forgotten that the discovery had been made before. From now on

The question to be examined was: how exactly did they orbit the sun? On a circle centered on the sun, or along some other curve? How fast were they moving?
etc. These took longer to reply. The post-Copernican epochs were the periods when the questions were debated as to whether the planets really revolved around the sun with the earth or whether the earth was the center of the universe. Later Tycho Brahe[2] A man named suggested a method of answering the question. If the planets were observed very closely and their exact positions in the sky recorded, the state of the theories could perhaps be clarified. This was the key to modern science and the beginning of a true understanding of nature: observing something, recording details and hoping that this information contains clues that allow one interpretation or another to be drawn. Tycho, a wealthy man, had an island near Copenhagen. Here he placed huge circles of brass and built special observation posts. Then during the night he recorded the positions of the planets and only through such arduous and hard work can anything be found.

All information collected is Kepler's[3] hand over; He set out to study what kind of motion the planets made around the sun. He used the trial and error method. Eventually he thought he had the answer: the planets moved in circles with the sun as the center. But then he realized that one planet, I think Mars, deviates in an eight-minute arc. Kepler thought Tycho Brahe could not have made such a mistake and concluded that the answer was incorrect. Since the experiments had been done very carefully, he tried another way and finally discovered three things.

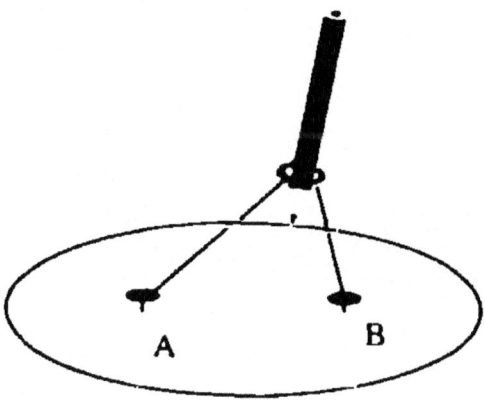

illustration 1

At first, the planets followed an elliptical orbit with the sun in focus. All points A and B on the ellipse are the focus. A planet's orbit around the sun is an ellipse; The sun is also in one of the focal points. The next question was: does its speed increase as it gets closer to the sun and slow down as it gets farther away? Kepler found the answer to this as well (Figure 2).

figure 2

His answer can be explained as follows: For example, let's determine the position of the planet at two different points in time a certain distance apart, about three weeks. Then, in another part of the orbit, let's determine two different positions of the planet, again three weeks apart, and draw the lines connecting the Sun and the planet (in scientific terms, these are:

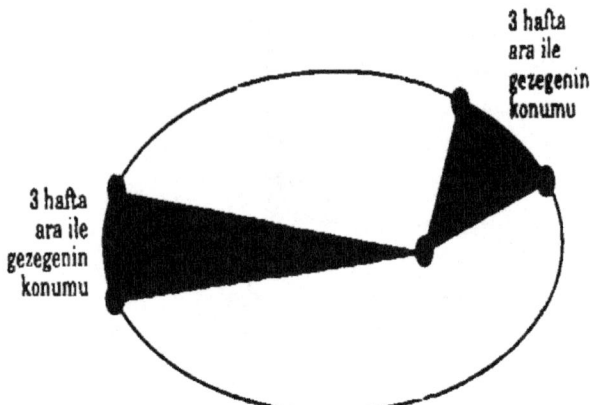

are radius vectors). The area between the two lines drawn at three-week intervals and the orbit is the same for each area of the orbit. So the planet moves faster when it's closer to the Sun and slower to scan the same area when it's farther away.

A few years later, Kepler discovered a third rule. This rule was not just about the movement of a single planet around the sun; He also established relationships between different planets. According to this rule, the time it takes for a planet to complete one revolution around the sun depends on the size of its orbit; this time is proportional to the square root of the third cubic meter of orbit. The size of the orbit is the largest diameter of the ellipse. Kepler's three laws can be summarized as follows: the orbit is an ellipse; equal areas are sampled in equal times, and the time required for one cycle is proportional to the three-second power of the dimension; ie by the square root of the cube of dimension.

The next question was: what moves the planets around the sun? Some of Kepler's contemporaries answered this question as follows: Angels flap their wings to push planets into orbit from behind. As you will see later, this answer is not far from the truth. The only difference is that the angels are seated facing different directions, flapping their wings inward.

At the same time, Galileo also studied the laws of motion of ordinary objects on Earth, and during this study he conducted some experiments. How did balls roll down an inclined plane, how did pendulums swing? etc. Galileo discovered an important rule called the principle of inertia. The rule was: an object moving at a certain speed in a straight line will move along that line at the same speed forever if nothing is done. roll a ball nonstop

Although it is hard to believe for anyone who has ever worked; Given these ideal conditions, friction etc. on the ground. without factors, the ball would actually fly at a steady speed forever.

The next development began with Newton's discussion of the question: What if the object doesn't move in a straight line? His answer was this: It takes power to change speed in any way. For example, if a ball is pushed in the direction it is moving, its speed will increase. When the direction of travel has changed, the force is applied from the side. Force can be measured by the product of two effects. How much the speed changes in a small time interval is defined as "acceleration". If we multiply this by the object's mass or the coefficient of inertia, we get the force. That is measurable. For example, if we turn a stone attached to the end of a string in a circle above our head,

The reason is this: although the speed of the stone is constant, as it spins in a circle, its direction changes, so a force is required to constantly pull the stone inward; This force is proportional to the mass. Now let's take two separate stones and rotate one, then the other and measure the force required for the second stone. This force will be greater than the first in proportion to the difference in their masses. Determining the force required to change velocity also creates a method of measuring mass. Newton drew a different conclusion from this. Let's explain it with a simple example: when a planet orbits the sun, it doesn't take any force to move sideways along the tangent. If there had been no violence he would have grabbed his head and left. But the planet doesn't do this; Without the force, it's not far from where it would have disappeared after a while, but close to the Sun (Figure 3). In other words, its speed and motion diverge in the direction of the sun; That is, angels must constantly flap their wings towards the sun.

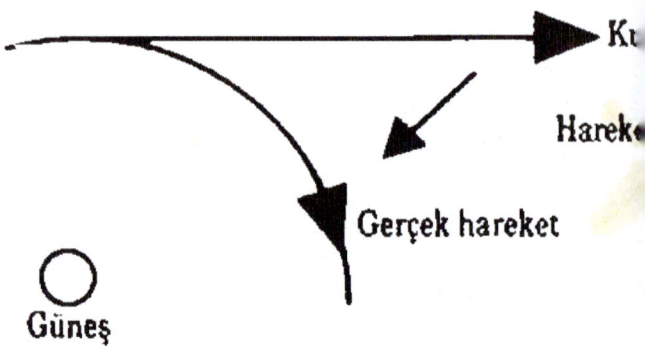

figure 3

There is no known reason why a planet should move in a straight line. The reason why objects keep running forever was not found. The theory of inertia also has no known origin. Although angels are not real, the fact is that the movement continues. However, force is required for the phenomenon of falling, and it is understood that the origin of force is in the direction of the sun. Newton's theory of sweeps of equal areas in equal times is a direct result of the assertion that all changes in velocity are in the direction of the sun; succeeded in showing that this also applies to elliptical orbits. I will explain this in more detail in my next presentation.

Using this law, Newton showed that the force acts in the direction of the Sun, and by knowing how the periods of the planets vary with their distance from the Sun, one can also determine how this force varies with distance, and determines that the force is inversely proportional to the square of the distance.

So far Newton hasn't said much; because he was merely expressing two different things that Kepler meant. First, by saying that the force is directed toward the sun; Second, it was equivalent to saying that force is inversely proportional to distance squared.

People had seen Jupiter's moons move around Jupiter through a telescope. This movement was just like in the solar system; as if the moons were pointing towards Jupiter they were drawn. The moon is also in Earth's gravity; The earth rotates and is pulled towards the earth. His next theory is that everything seems to be under the attraction of the other; If we make a generalization, this led to the conclusion that every object attracts every object.

If this were true, the earth would pull the moon towards it just as the sun pulls the planets. It's a well-known fact that the earth attracts objects - we all know that we're stuck in our chairs even though we want to fly through the air. Gravity on Earth is something we know well as a gravitational phenomenon. Newton thought that the force of gravity holding the moon in orbit might be the same as the force pulling objects toward Earth.

It's not hard to figure out how far the moon moves in a second. We know the size of the orbit; We also know that the moon orbits the earth in one month. So we can calculate how far I travel in one second. From here, how far below the straight line the Moon is can be calculated by going along the orbit, not along the straight line. This distance is one inch[4] one twentieth (0.127 cm). The moon's distance from the center of the earth is sixty times our distance from the center of the earth. We are from the middle 4,000 miles[5] away, the moon is 240,000 miles from the center of the earth. In this case, if the inverse square law is correct, an object on Earth falls at 1/20 inch x 3,600 (60 square) per second. Because until the force goes to A/a, according to the inverse square law, it weakens 60 x 60 times. 1/20 inch x 3,600 is approximately 16 feet (feet).[6] From Galileo's experiments, it was known that objects fall at a speed of 16 feet per second. This showed that Newton was on the right track and there was no turning back. Because it's a brand new concept that combines two completely unrelated phenomena, like the moon's orbital period and its distance from Earth, and how far an object on Earth would travel in one second if it fell.

something was found. This result was a stunning test that showed everything was right.

Later, Newton came up with many new things. He calculated what the shape of the orbit would be if the law of gravitation were inverse square and found that it was an ellipse. Also, many different events were explained. One of these was the ebb and flow event. The tides were caused by the gravitational pull of the earth and seas by the moon. This has been taken into account previously; However, there was a problem: if the event were due to the moon pulling the seas, the water on the moon's side would rise, then there would only be one high tide per day (Figure 4). In fact, we know that there are two tides a day, roughly every twelve hours.

figure 4

There was another school of thought that came to a different conclusion. Accordingly, the earth was pulled out of the water by the moon. Newton was the first to notice what was really going on: the moon's gravitational pull on land and sea at the same distance was equal. The waters at y are closer to the moon than Earth, and the waters at x are farther away. The earth is solid, not liquid. Therefore, the water at y is drawn more to the moon than to earth, and the water at x is drawn less to the moon. As can be seen from a sort of combination of these two images, two ebb and flow events occur. In reality, the earth also moves in a circular orbit like the moon. The force exerted by the moon on the earth is balanced; but what is the stabilizer? The earth moves in a circular orbit, just like the moon moves in a circular orbit to balance the earth's gravitational pull. The center of this circle is at a point inside the earth and makes a circular motion to balance the power of the moon. Since both rotate about a common center, the forces for the earth are balanced; However, since less water is drawn into x and more water into y, the water swells on both sides. Anyway, that explained the tides and the reasons why it happened twice a day. Many more have since been revealed: the earth was round because everything was being drawn inward; It wasn't round because it rotated on its own axis. The outer zones were shifted a little further and an equilibrium was established. However, that explained the tides and the reasons why it happened twice a day. Many more have since been revealed: the earth was round because everything was being drawn inward; It wasn't round because it rotated on its own axis. The outer zones were shifted a little further and an equilibrium was established. Anyway, that explained the tides and the reasons why it happened twice a day. Many more have since been revealed: the earth was round because everything was being drawn inward; It wasn't round because it rotated on its own axis. The outer zones were shifted a little further and an equilibrium was established. because it rotated on its own axis. The outer zones were shifted a little further and an equilibrium was established. Anyway, that explained the tides and the reasons why it happened twice a day. Many more have since been revealed: the earth was round because everything was being drawn inward; It wasn't round because it rotated on its own axis. The outer zones were shifted a little further and an equilibrium was established. because it rotated on its own axis. The outer zones were shifted a little further and an equilibrium was established. Anyway, that explained the tides and the reasons why it happened twice a day. Many more have since been revealed: the earth was round because everything was being drawn inward; It wasn't round because it rotated on its own axis. The outer zones were shifted a little further and an equilibrium was established.

As science advanced and more accurate measurements were made, Newton's law faced greater challenges. The first of these dealt with the moons of Jupiter. With careful observation over a long period of time, it was possible to determine that their motion conformed to Newton's law. However, the result showed that this was not true. Jupiter's moons deviated from the time calculated by Newton's law, sometimes eight minutes ahead, sometimes eight minutes behind. This difference was forward when Jupiter was close to Earth and backward when it was farther away. This was a strange situation. Roman who has full faith in the law of gravity[7], in this case,

He came to an interesting conclusion that it takes time for light from Jupiter's moons to reach Earth. Also, when we looked at these moons, we didn't see their current state, but their state before the time it took for the light to reach us. When Jupiter is close to us, the light arrives in less time, and when it's far away, it takes longer. Therefore, Roemer had to correct the observations for the time shift as early or as late. In this way he was able to measure the speed of light and to show for the first time that light does not propagate instantaneously.

I want to draw special attention to this point: if one law is true, it can lead to the discovery of another law. If we trust one law, the occurrence of something to the contrary leads us to another fact. Nor would we know what to expect from Jupiter's moons if we did not know the law of gravity; The speed of light would have been measured much later. This process led to an era of discovery. Each new discovery brings with it tools that lead to another. So this age, which has been going on for four hundred years and will continue to go on at a great pace, began in this way.

A new problem later emerged. According to Newton's law, the planets were not only in the gravity of the sun; They also got dressed a little. So their orbits should not be elliptical. It was a small shot, however; but the "little ones" can also be important and influence the movement. Jupiter, Saturn and Uranus were known as major planets. Calculations and observations were performed to determine the extent to which their orbits differed from Kepler's perfect ellipses due to gravitational pull. As a result, Jupiter and Saturn acted according to the calculations; It

turned out that Uranus was acting strangely. Another opportunity to show that Newton's laws are wrong... But brace yourself! Adams and Leverrier[8th] called two people
As a result of their independent studies, they almost simultaneously suggested that the movements of Uranus were influenced by an invisible planet. Each of them said to his observatory: "Turn your telescope and observe. You will see a new planet," they sent a letter. The reaction of one of the observatories was "Nonsense! "Someone sitting with pen and paper in hand tells us where to look to find a planet." The other observatory was more...well, its method was different, and it found Neptune.

More recently, at the beginning of the 20th century, it was recognized that the movement of Mercury is also not quite "correct". This caused a lot of trouble until Einstein showed that Newton's laws were somewhat flawed and needed to be changed.

Now the question arises as to the breadth of the scope of this law. Does the law also apply outside the solar system? I am now going to show you in Figure 1 the evidence that the law of gravitation extends beyond the solar system. Here we see three separate photos of what we call a double star. There happens to be a third star in the picture. This tells us that someone did not rotate the frame, as is easily possible in astronomy, but that the binary actually does. The stars do turn; You can see the trajectory they followed in Figure 5. It is clear that they attract each other and, as expected, rotate along an ellipse. The points in the figure are positions that they follow sequentially in a clockwise direction at different times.

Figure 5

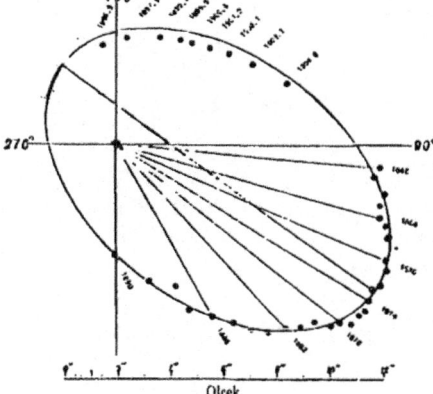

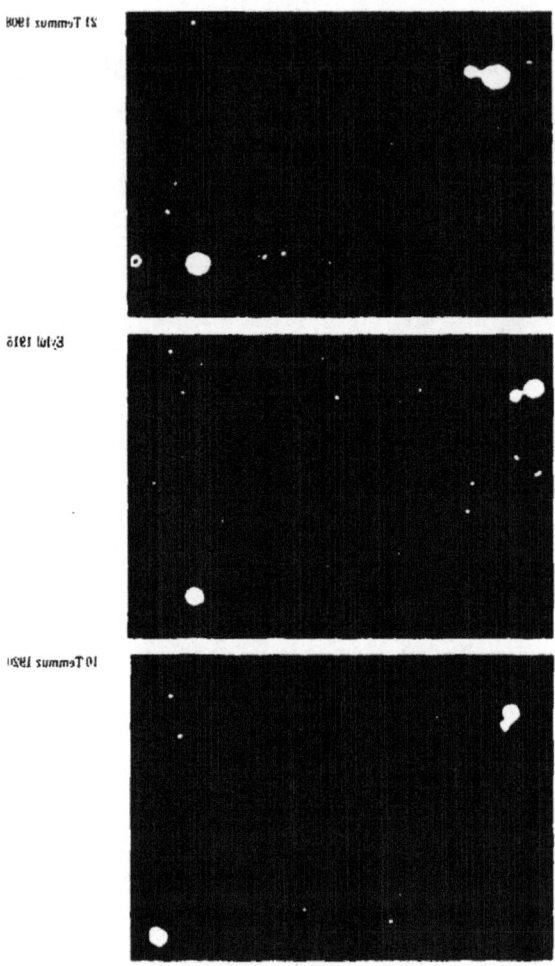

Figure 2 A globular cluster

Figure 3 spiral galaxy

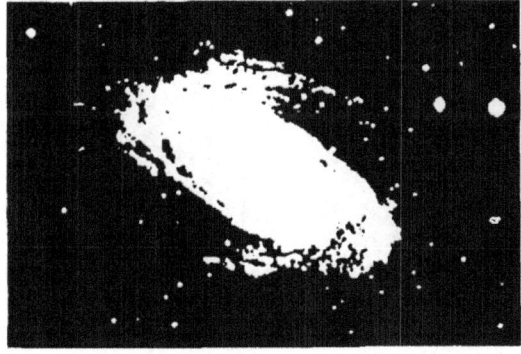

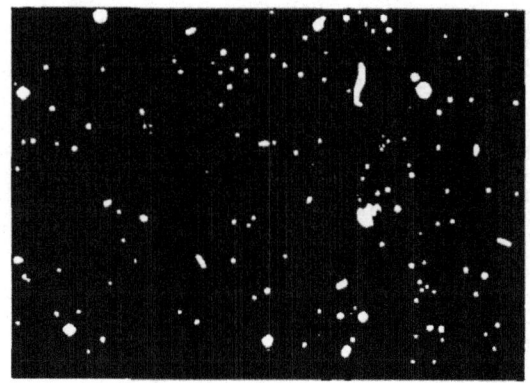

Figure 4 A galaxy cluster

Figure 6 A gas nebula

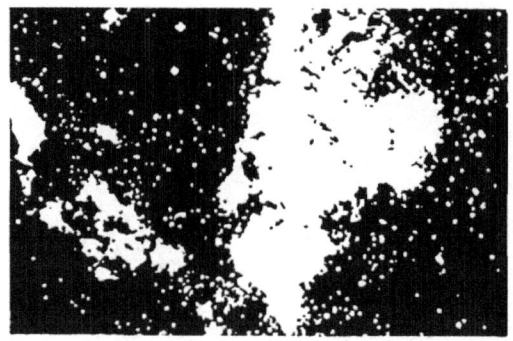

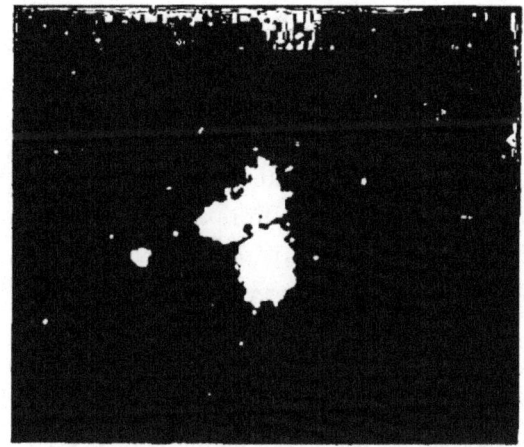

1947

1954

Figure 6 Evidence of new star formation

Everything is there; However, in case you haven't noticed, I warn you that the center is not at the focus of the ellipse, but rather far away. Is there something wrong with the law? No, God is showing us this trajectory from a different angle, not straight ahead. After marking the focus of an ellipse, if you hold the paper at any angle and look at it projectively, you'll see that the focus may not match its projection. It looks like this because the orbit is tilted in space.

What can we say about longer distances? Does this force exist between two stars beyond two to three times the diameter of the solar system? In Figure 2, a cluster of stars with a very large number of stars 100,000 times the diameter of the Solar System can be seen. The white spot in the image is not a single white spot, but it appears so because the optical devices have insufficient discrimination. In fact, in this large globular cluster there are tiny dots that move up and down, back and forth, far apart, just like other stars that don't collide. This is one of the most beautiful things to see in the sky. As beautiful as the waves of the sea and the sunset... What holds the galaxy together is the gravitational pull between the stars. The distribution of matter and the magnitude of the distance, It allows us to roughly figure out what the law of gravitational force between stars is, and the result is the inverse square law. It is impossible for the sensitivity of these measurements and calculations to come close to that of the solar system.

The gravitational force goes even further. The star cluster in Figure 2 is the size of a pinhead in a typical large galaxy shown in Figure 3. It's obvious that this is held together by a force, and the next candidate that comes to mind is the gravitational force. We cannot measure the validity of the inverse square rule for such large dimensions. earth to sun
These galaxies are 50,000 to 100,000 light-years long, although the distance between them is eight light-minutes. However, there is no reason to doubt that the gravitational pull in these large star clusters persists at such distances. Image 4 is proof that this power goes further. A cluster of galaxies can be seen in this image. The galaxies formed a star-cluster-like cluster; but the things that are piling up here are the "big babies" in picture 3.

So far we can prove directly that gravitational force exists; that is, a distance a tenth or a hundredth the size of the universe. Accordingly, although you read news in the newspapers about something being outside of Earth's gravitational pull, there is no definite end to Earth's gravity. This gravity becomes weaker and weaker inversely than the square of the distance; if the distance is doubled, it becomes weaker by a factor of four and is thus lost in the turmoil of the strong fields of other stars. It creates a galaxy by attracting other stars along with the surrounding stars, which in turn attract other galaxies, forming a cluster of galaxies. Thus, the earth's gravitational field never ends;

The law of attraction is different from most other laws. It is clear that it is very important to the economy and mechanism of the universe, and it has many practical applications related to the universe. However, it has an atypical property that differs from other physical laws: its knowledge is of little practical use. By choosing the law of gravitation as an example of physical laws, I am choosing an example with an atypical feature. However, I should add that it is impossible to choose something that is not in some way atypical by choosing only one thing among things. This is one of the mysteries of the world. The law of gravity, as far as I know, in geological mineral exploration; in forecasting tides;

in a more modern calculation of the movements of outgoing satellite and planetary probes; and it is used to calculate the positions of the planets in a more modern way, which is necessary for those who write horoscopes in magazines. The world we live in is truly an incredible world; Advances in science only serve to continue 2,000 years of nonsense.

I should also mention important things where the law of attraction really has an impact on the behavior of the universe. One of the most interesting of these is the formation of new stars. In picture 5 we see a gas nebula that is located in our galaxy. The galaxy is all gas, not many stars. Black spots are places where gas has been compressed or drawn in. Maybe it was a shock wave that started it all. The next events are the gathering of the gas by the action of the gravitational force, the formation of large heaps and balls of gas and dust. As they fall inward, they burn in the heat of the fall and become stars. Figure 6 shows some evidence of new star formation.

This is how stars form when the gas is compressed and brought together by gravity. Stars sometimes spit out dust and gas when they explode, which regroups and creates new stars (a process reminiscent of a recirculation engine).

I showed earlier that gravity also exists at great distances. But Newton said that everything attracts everything, do two bodies really attract? Can we do a direct experiment instead of waiting to see if the planets attract each other?

Such an experiment, Cavendish[9] was fabricated using the material shown in Figure 6. According to this, a rod with spherical masses at both ends hangs from the end of a very, very thin quartz wire.

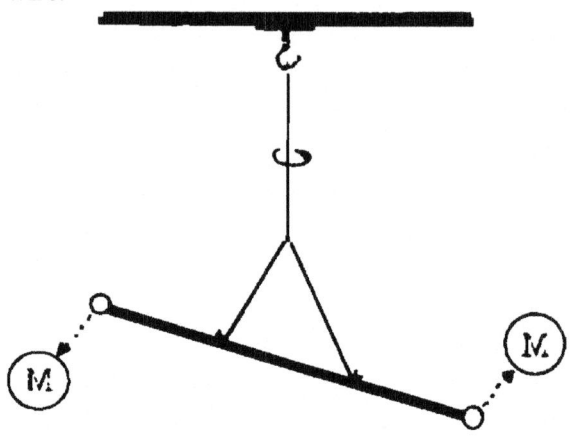

Figure 6

Then, as shown in the figure, two large balls of lead are placed on the sides of the masses. The attraction between the balls causes a small bend in the wire. The normal force of attraction between objects is very small, it is possible to measure this force between two spheres. Cavendish called this experiment the "weighing of the world". As a result of the wise and careful training now practiced, we instead talk to our students about "Measuring the Mass of the Earth." Cavendish used a direct method to measure force, two masses, and the distance between them; so he was able to find the gravitational constant G. Now you are "We're in the same situation again; We know the drag force, the mass of the object being dragged, and the distance between them.

This experiment is indirectly the first determination of how big and how heavy the ball is that we are standing on. so much
It was an amazing feat, and I think that's why Cavendish called his experiment "weighing the earth" rather than "calculating the constant in the gravitational equation." It also weighed the sun and everything else. Because the same method applies to the gravitational force of the sun.

Another question about the law of attraction is whether gravity is really proportional to mass. That is, gravitational force is strictly proportional to mass; Response to force, motion as a result of force, and changes in velocity that are inversely proportional to mass.

This means that the velocities of two bodies of different masses change in the same way in the gravitational field; or two objects in a ventilated environment fall to the ground in the same way, regardless of their mass. This is Galileo's famous Leaning Tower of Pisa experiment. To explain with an example: an object inside an artificial satellite orbits the earth in the same orbit as an object outside the satellite; It will be like floating in the air. The fact that force is directly proportional to mass and the effects are inversely proportional brings about this interesting result.

What is the sensitivity? This point was made by Eötvös in 1809.[10] lately with greater certainty thickness[11.]and it was found to be one in 10,000,000,000. Forces are exactly proportional to mass. How can such sensitive measurements be performed? Suppose you want to measure whether the "gravity of the sun" measurement is correct. You know that the sun draws us all and of course the earth to itself. However, you want to know if this attraction is exactly proportional to inertia. The experiment was carried out first with sandalwood, then with lead and copper; Today polyethylene is used. The earth revolves around the sun and objects are thrown outwards due to inertia. This ejection occurs in proportion to the inertia of the two objects. However, according to the law of gravitation, these two bodies are directed towards the sun in proportion to their mass. You are drawn correctly. If their pull towards the sun differs from their ejection due to their inertia, one will be pulled towards the sun while the other will be repelled from the sun. When we attach these objects to either end of the rod attached to Cavendish's quartz wire, the wire bends toward the sun. However, the wire shows no such bend. So the gravity exerted by the sun on these two bodies is exactly proportional to the centrifugal effect, which we call inertia. Therefore, the gravitational force exerted on an object is exactly proportional to the object's coefficient of inertia, i.e. its mass.

One thing is particularly interesting. The inverse square law appears elsewhere; For example, electric charges exert forces that are inversely proportional to the square of the distance between them. This suggests that the inverse of the square of the distance may have a very profound meaning. But nobody has been able to show that electricity and gravity are different aspects of one thing. Today the theories of physics are a collection of parts and parts whose physical laws are not completely compatible with each other. No single structure was found from which everything can be logically deduced. We just have a large number of parts that don't quite fit together. So at these conferences I don't have to talk about what the law of physics is, but about what different laws have in common. We don't know the connection between them. However, it is very surprising that some things are the same in these two laws. Let's look again at the Electricity Act.

The force is inversely proportional to the

square of the distance; What is interesting, however, is the enormous difference between the strengths of electricity and gravitational forces. Anyone wanting to harvest electricity and gravity from a common structure will find that electricity is much stronger than gravity. It is hard to believe that these two can come from the same origin. How can you say that one thing is stronger than the other?

This difference depends on how much charge and how much mass there is. You can't say "If I get a lump this size" to show that gravity is stronger; because you determine the size. If we want to look at something that nature produces—nature's pure numbers have nothing to do with centimeters, years, or anything of our dimensions—we can do it like this: electron-different particles make different numbers; To give an example, let's choose the electron - we choose an elementary particle like, two electrons are two elementary particles; Because of the electricity, they repel each other in inverse proportion to the square of the distance.

Two incredibly disproportionate forces - one pulling and one pushing - what kind of equation could the solution be?

People tried to find other things where the relationship between them could be that big. If such a large number is to be sought, one could consider, for example, the ratio between the diameter of the universe and the diameter of a proton. It is amazing that it is also a 42-digit number. An interesting thesis was then put forward that this ratio is the ratio between the size of the universe and the size of the proton. But the universe expands over time; so the gravitational constant also changes with time. This is a possibility, but no evidence has been found to prove it's real. There is some evidence that the gravitational coefficient does not change in this way.

Before we close the subject of the theory of gravitation, two things need to be mentioned. First, Einstein had to change the laws of gravitation according to his own principles of relativity. The first of the principles stated that 'x' could not occur at once. However, Newton's laws stated that the force occurred in an instant. Einstein had to change Newton's laws. However, these changes had little impact. One is that all masses fall, light contains energy, and energy equals mass. The light falls accordingly; which means light near the sun is deflected, which is actually deflected. The gravitational force in Einstein's theory is also slightly modified. So the law to a very small extent; It has changed enough

Eventually, a change in the physical laws governing small scales was required; it was discovered that matter on a small scale is subject to completely different laws than on a large scale. Then the question arises: how does the law of gravitation hold on a small scale? This is the quantum of gravity.
We call it theory. There is still no quantum theory of gravity. We have not been able to find a theory that is compatible with both the uncertainty principles and the principles of quantum mechanics.

Now you're going to say to me, 'Yes, you told us what happened. But what is gravity? Where does it come from? Are you saying that the planet looked at the sun and saw how far away it was, squared it, and then decided to act according to the law?" Let me put it another way. I have expressed the law mathematically; but I said nothing about the mechanism. In my next talk I will talk about how this can be done; namely from "The Relation between Mathematics and Physics".

At the end of this talk, I would like to

highlight what the Law of Attraction has in common with some of the other laws mentioned: First, the way it is expressed is mathematical; so are the others. Second, it's not entirely true. Einstein had to change it; However, we know it's not entirely true. Because it does not yet cover quantum theory in its current form. These also apply to all of our other laws; none is quite correct. There is a limit that is always mysterious, there is always something to do. This may or may not be a feature of nature; However, it is a feature common to all laws we know today. Maybe it's just a lack of knowledge.

    The most striking feature of the Law of Attraction is its simplicity. its principles are easy to formulate precisely; there is no ambiguity that requires a change in their essence. It's simple, so it's beautiful. I'm not saying its effects are easy; The motions of the planets and the perturbations they exert on each other require calculations that can be very complex. Detecting the movements of stars in a globular cluster is beyond our capabilities.

The effects are complicated; However, the main model or underlying system of all is simple. This is a common feature of all laws. The real implications are complex, but they are simple.

    Finally, I will address the universality of the law of attraction and its validity over very large distances. Newton was able to predict that Cavendish's miniature model of the solar system, which represents the solar system, i.e. the attraction between the two spheres, would also hold true in the solar system, which would be obtained at a magnification of a hundred trillion times. Then we see that galaxies a hundred trillion times this size also attract by the same law. Because nature weaves her patterns with only the longest threads, each small part of the weaving reveals the layout of the entire rug.

The relationship between mathematics and physics

Looking at the applications of math and physics, it is obvious that math will come in handy in complex situations involving large numbers. For example, in biology, the question of the effect of a virus on a bacterium cannot be mathematical. When we look through a microscope, a tiny, flickering virus can transform into an oddly shaped bacterium. -They all have different shapes- You will see that they catch it from one point. Maybe you gave him his DNA, maybe you didn't. But if we do this experiment with millions of bacteria and

viruses, we can learn a lot about viruses through averaging. To determine whether viruses develop in bacteria and what species occur with what frequency.
takes mathematical mean and hence genetics, mutation etc. We can explore themes.

As a small example, consider a giant chessboard on which chess or checkers is played. The realization of a single movement is not mathematical. In a game played with many pieces on a huge board, analyzing the best moves, or analyzing good and bad moves, requires thorough reasoning. This means that someone has thought about it beforehand. Here's a type of math that involves abstract thinking. Another example is turning computers on and off.

If there's just a switch that can be turned on or off, there's nothing mathematical about it, although mathematicians like to confuse it with their mathematics. But it takes math to know what's going on in a very large system of interconnected wires.

I would like to point out right away that in complex situations, when the basic rules of the game are known, mathematics has a huge area of application in physics for studying events in detail. If our topic were simply the relationship between math and physics, I would spend most of my time explaining it. But since this is only part of a series of talks about the properties of physical laws and there is no time to think about what happens in complex situations, I will now move on to another topic, the properties of fundamental laws.

If we consider the game of drafts, the Basic Laws cover the rules by which the pieces should be played. In a complex situation, mathematics can be applied to figure out what would be a good move under certain conditions; However, the simple nature of the basic rules requires very little mathematics. Rules are easily expressed in colloquial language.

The difference to physics is that mathematics is required for the basic laws. I will give two examples; one doesn't really need mathematics, the other does. According to Faraday's physical law, the amount of substance released during electrolysis is determined by the current and the time it takes for the current to pass through; That is, the amount of matter released is proportional to the electrical charge flowing through the system. An explanation that seems very mathematical... But what actually happens is this: every electron that passes through the wire carries an electric charge. As a specific example, let's assume that a single electron has to pass for an atom of matter. Then the number of atoms released must be equal to the number of electrons passing the wire; So it is proportional to the electric charge flowing through the wire. Basically, this seemingly mathematical law does not contain anything very deep and does not require any mathematical knowledge. The fact that one electron is needed for each atom is perhaps mathematical; but what I'm talking about here isn't that kind of math.

Now let's consider Newton's law of gravitation, the properties of which I have already mentioned. To emphasize the speed at which mathematical symbols convey information, I wrote you the following equation:

$$F = G \frac{mm'}{r^2}$$

The force is proportional to the product of the masses of the two objects, inversely proportional to the square of the distance and the objects' response to the force by changing velocity; or that they show their motion by changing it in the direction of the force, directly proportional to the force and inversely proportional to the masses. These are words;

You don't have to write equations. Still, they are mathematical in a way, and why is there a fundamental law?

We wonder what they created. What is the planet doing? Does it face the sun, see how far it is, and use an internal calculator to inverse the square of the distance and figure out how far it needs to travel? This cannot be the explanation for the gravitational mechanism. You might want to know more; a lot of people wanted it. First, Newton was asked about his theory: "But that doesn't mean anything; That doesn't tell us anything." He also says, "He tells you how he moves; that should be enough. And I told you how he acted, not why he acted the way he did it." However, people are not always satisfied if they don't know the basics of the job.I will pick out an example of the theories that have been constructed to give you the kind of answer

Suppose many particles fly all over the earth with great speed. They come evenly from all directions and occasionally nudge us. We and the sun are almost transparent to them; but almost, so not quite. Some come and crash into us. Let's see what happens (Figure 8).

Figure 8

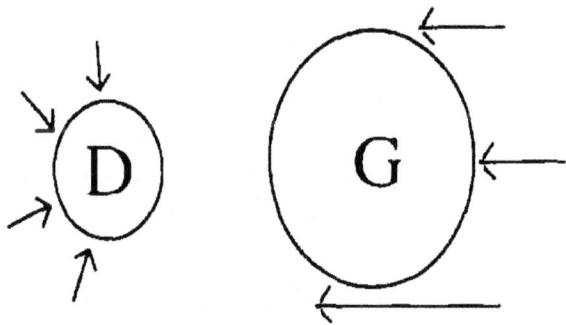

D denotes earth and G denotes sun. If the sun weren't there, the particles would bombard the earth from all directions;

The shattered particles would cause slight rattles and clicks. Since these thrusters came from all directions, they would not shake the earth in any direction. However, since the sun is there, particles coming from that direction cannot come to earth because they hit the sun and are partially absorbed by the sun. Therefore, those who come to earth from the direction of the sun will be fewer than those who come from other directions because they encounter an obstacle, namely the sun. The further away the sun is, the smaller the proportion of particles that come from all directions. The sun will actually appear smaller in proportion to the inverse square. For these reasons, a repulsive force to the sun occurs on earth, which is proportional to the reciprocal of the square of the distance. This power is the result of a large number of very simple effects, one after the other of impacts. When these are taken into account, the oddness of the mathematical context is greatly reduced. Because the basic function is much simpler than calculating the square of the reciprocal of the distance. The order in which the particles jump makes the calculations.

The only flaw with this statement, however, is that it is not valid for a number of reasons. Any theory we propose must be examined for all of its possible consequences in order to understand what else it predicts. There is something else that this theory boils down to. As the earth moves, more particles hit the front than the back (if you walk in the rain, more rain hits your face than the back of your head because you're running against the rain). Likewise, as the earth moves, it moves into the particles pointing towards it and away from the particles behind it. As more particles hit the front from behind, there is a force opposing the motion. This force slows the Earth's motion in its orbit. At that point, the Earth would not have stayed in orbit around the Sun for (at least) three or four billion years. That means the end of the theory. Now he tells me, 'Well, it was a good theory; me for a while saved from math. Maybe I can find a better one," you might say. You might find; because nobody knows the exact result. However, since Newton, no one has managed to find a theoretical explanation that can replace the mathematical mechanism behind this law that does not express the same thing, does not complicate the mathematics, or whose results do not contradict specific events, has not become a non-mathematical model for the law of gravitation found.

If this was the only law with this function it would be interesting but a bit annoying. However, the truth is that the more we researched, the more laws we discovered and the deeper we looked into nature, the more common this disease turned out to be. Each of our laws is a mathematical expression involving highly complex and intricate mathematics. Expressing Newton's law of gravitation is relatively simple mathematics. The further you go, the more difficult and complex the expressions become. Why? I have no idea. I just want to draw your attention to this situation. This is the focus of my talk: I want to emphasize that the beauties of nature cannot be explained in a way that can be felt by those who do not have a deep understanding of mathematics. Sorry, but that seems to be the case.

You might say, "Well, if the law doesn't have an explanation, at least explain what the law is. Can't you describe it with words instead of symbols? Mathematics is just a language; I would like to be able to translate it." With a little patience I can do it; I've already done some of it. I would go a little further and give a more detailed explanation, I would say that the equation means that if the distance is doubled the force is a quarter. I can also convert all symbols into words. That means I can be more considerate of ordinary people who sit and wait for me to explain. These abstract and complex laws are commonplace.

People who have tried to explain people in the language of an ordinary person have a very different reputation. The common man will try these books one by one and try to find a book that does not contain the difficulties he will definitely encounter in the end. As he reads he will encounter increasing confusion, a succession of muddled statements, things that seem incoherent and difficult to understand. Perhaps he is clinging to another book in hopes of finding an explanation. This author almost succeeded, maybe someone else really does ...

However, I don't think this is possible. Because mathematics is not just another language, it is a language plus reasoning;

something like a language plus logic, a reasoning tool. More specifically, it is an accumulation resulting from a person's careful thought and consideration. Thanks to mathematics, a connection can be made between one statement and another. For example, I could say that the force is directed towards the sun. Also, as I said before, if I draw a line between the sun and the planet and after a certain time, like about three weeks, draw another line; the area that will be swept by the planet three weeks later, the next three weeks after that, etc. I can say it's moving in the same way as the areas which it will scan. I can carefully explain what I said in these two statements; but I can't explain why the two are the same. Nature, its enormous intricacies revealed through a series of strange rules and laws, all carefully explained, are indeed very much intertwined. If you don't want mathematics, you won't be able to understand that it is only possible to go from one phenomenon to another with logic.

It may seem incredible to you to show that equal areas are sampled at equal time intervals when the forces are directed towards the Sun, please excuse me I will now explain to you how we can prove that these two things are equivalent. So you understand more than the letter of the law. Don't just argue that the two laws are linked.

I will show that you can pull through and that math is just organized thinking. Then you will also see the beauty of the relationship between the expressions of the laws. I will now prove to you that when the forces are directed towards the sun, equal areas are scanned in equal times.
Common height of triangles

Figure 9

Consider a sun and a planet (Figure 9) and assume that the planet is in position 1 at some point in time. After one second, it moves to position 2. When the sun is not exerting any force on the planet, according to Galileo's principle of inertia, the planet moves along the straight line without interruption. So he drives in the same time interval, i.e. in the next second, on

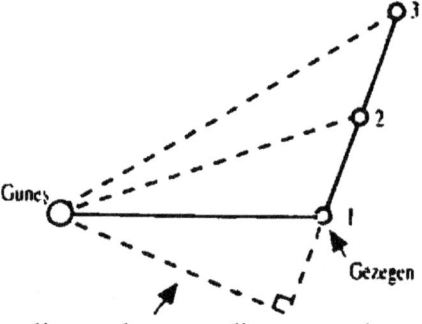

the same line at the same distance and reaches the 3rd position. We will first show that equal areas are sampled when no force is acting. Let's remember that the area of a triangle is the base multiplied by the height, and the height is the perpendicular distance from the base. If the triangle is obtuse (Figure 10), AD is its perpendicular distance and BC is its base.

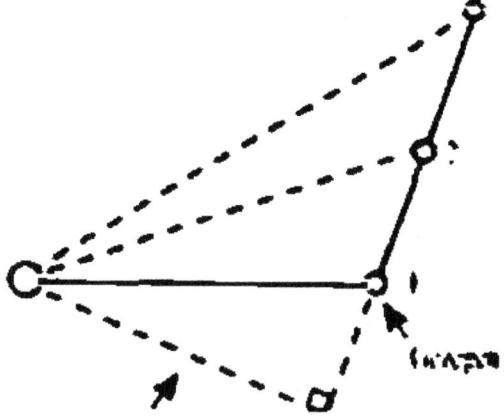

Figure 10

Now let's compare the areas to be scanned when the sun is not exerting any force (Figure 9). If you remember, intervals 1-2 and 3-4 were the same. Are the areas the same? Consider the triangle formed by the Sun and points 1 and 2. What is the area of this triangle? The area is the base 1-2 times the perpendicular distance from the apex S to the base. What can we say about the other triangle we get when we go from 2 to 3? Its area is also equal to the product of the base 2-3 times half the perpendicular distance to S. The altitudes of the two triangles are equal; as said, their basics are the same; hence their ranges are equal. So far everything ok. If the sun were not exerting any force, equal areas would be swept out in equal times. However, there is a power that comes from the sun. In the 1-2-3 range, the Sun applies gravity and always attracts movement in different directions towards itself.

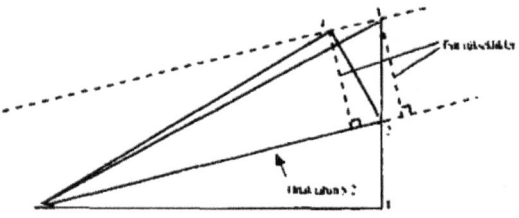

Figure 11

We can explain it like this: although the masses are moving on the 1-2 line, and in the next second they would be moving on the same line if there were no force, the motion changed to a direction parallel to the 2-S- line the effect of the sun. The movement that follows will be a combination of what the planet wants to do and the change brought about by the Sun's influence. So the planet moves to position 4 instead of position 3. Now we will compare the areas of triangles 23S and 24S and I will show you that they are equal. The bases of these triangles are equal: S-2 Are their heights equal? Certainly; because they lie between parallel lines. The distance from 4 to line S-2 is equal to the

distance from 3 to the extension of line S-2. The areas of the triangles S24 and S23 are therefore equal. I have already proved that fields S12 and S23 are equal; then, we know that S12 = S24. Thus, in the first second and second second of the planet's actual motion in its orbit, equal areas were sampled. Hence, by reasoning, we find the relation between the force directed to the sun and the equal areas. Isn't that easy?! I quoted Newton directly. Everything including the pictures is from Principia. Alone Hence, by reasoning, we find the relation between the force directed to the sun and the equal areas. Isn't that easy?! I quoted Newton directly. Everything including the pictures is from Principia. Alone Hence, by reasoning, we find the relation between the force directed to the sun and the equal areas. Isn't that easy?! I quoted Newton directly. Everything including the pictures is from Principia. Alone
the signs are different; He used Roman numerals, I used Arabic numerals.

All the proofs that Newton gives in his book are geometric. We no longer use this type of reasoning, we reason analytically using symbols. It takes creativity to draw right triangles, find their faces, and think about how to do it all. In the meantime, the analysis methods have improved, they have become faster and more efficient. Now I want to show you how it is with the symbols of mathematics; Operations require nothing but a set of symbols.

I want to talk about how fast the field is changing; I also refer to it as Â. As the radius rotates, the area changes, and how fast the area changes is calculated as the radius multiplied by the component of the velocity in the radius direction. This is the radial distance component multiplied by the velocity or rate of change of distance.

$$\dot{A} = \vec{r} \times \dot{\vec{r}}$$

Now the question is whether the rate of change of the field itself changes, the principle is that the rate of change of the field is constant. We take the derivative of this again; That is, we play a game in which we put some points in the required places. We have to learn this game that consists of a set of rules that have proven to be very useful in these tasks. We write like this:

$$\ddot{A} = \dot{\vec{r}} \times \dot{\vec{r}} + \vec{r} \times \ddot{\vec{r}} = \vec{r} \times \vec{F}/m$$

The first term tells us to take the component

of the velocity in the direction perpendicular to the velocity. This is null; Velocity has the same direction as itself, the second derivative is acceleration r, or the derivative of velocity is force divided by mass. This expression tells us that the rate at which the field changes is the resultant of the force in the direction perpendicular to the radius. However, if the force acts in the direction of the radius,

$$\dot{r} = \vec{F}/m = 0 \text{ veya} \ddot{A} = 0$$

Newton also said this; there is no force perpendicular to the radius; This means that the rate of change of the field is constant.

This shows us the power of an analysis with a different notation. Newton more or less knew how to do this with a slightly different notation. But he wrote his work geometrically for everyone to read. He invented calculus, the math I just used.

This example illustrates very well the relationship between mathematics and physics. When problems in physics become very difficult, we turn to mathematicians; They may have worked on such things and prepared in advance the mindset that we should follow. On the other hand, this may not have happened. Then we find our own train of thought and then pass it on to mathematicians. Anyone who has thought something through contributes to their knowledge of what is going on about it. If he summarizes them and sends them to the mathematics department, this information will be included in the books as a new subject of mathematics. So math is a way of moving from one statement to the next. The benefit for physics is obvious. Because there are different ways to talk about objects; improve us results in mathematics, It offers the possibility of making some changes in the law to analyze the situations and make connections between different expressions. In reality, physicists don't know much. It is enough for them to remember the rules that allow them to go from one point to another. For at the same times forces are radii direction, etc. All statements about him are linked by reasoning.

    Now an interesting question arises. Is there a position where we can deduce everything logically? Is there some pattern or pattern of nature that might enable us to understand that some propositions are fundamental and others consequential? There are two different perspectives on mathematics. In this lecture series I will refer to these as the Babylonian tradition and the ancient Greek tradition. In the Babylonian school of mathematics, the student learned by solving many examples until he realized the general rule. He was also pretty good at geometry; It was also necessary to have knowledge of various properties of circles, the Pythagorean theorem, area formulas of cubes and triangles. Added to this was the technique of reflection, which enabled him to to switch from one topic to another. Numeric quantity tables were also available to help him solve detailed equations. Everything needed for the calculations was ready. However, Euclid discovered a method by which all the theorems of geometry could be derived from a very simple set of axioms. In what I call Babylonian mathematics you know many theorems and most of the connections between them; but you don't quite realize that all of this can be deduced from a set of axioms. Recent mathematics focuses on what is and is not acceptable as axioms, with axioms and very strictly defined rules. Modern geometry shows how the whole system can be derived starting from the Euclidean axioms, which it slightly modifies, to make them more precise. For example, a theorem like the Pythagorean theorem (the sum of the areas of the squares drawn on the two sides of a right triangle equals the area of the square on the hypotenuse) cannot be expected as an axiom. On the other hand, from another point of view of geometry, from Descartes' point of view, the Pythagorean theorem is an axiom.

So we must first admit that we can also start from different points in mathematics. If all the theorems are related by reasoning, there is no reason to say that "these are the most fundamental axioms"; because if something else is said, one can argue differently. This is similar to a multi-part and multi-unit bridge; If some pieces fell off, you can reconnect them with a difference. The current mathematical approach is as follows; Construction of the structure based on the ideas chosen as axioms with a kind of preliminary contract. The Babylonian approach is: "I know this and I know this, maybe I know this too and I draw everything from them. I can forget tomorrow that this is true; but I remember something else is true and I'm rebuilding. I'm never quite sure where to start and where to stop. I just remember enough, always, enough. That way I can put them back together every day, even if my memory weakens and some pieces fall out. " It is not always efficient to derive propositions from axioms. When working on something in geometry, it is not efficient to start from the axioms every time. If you remember a few geometric things, you can always move on to other things; however, it would be more efficient to do the work in the opposite direction. Deciding which axioms are best may not be the most efficient way to solve the problem either. In physics, the Babylonian method is needed, not the Euclidean or Ancient Greek method. Why it is like that, I would like to explain. That way I can put them back together every day, even if my memory weakens and some pieces fall out. " It is not always efficient to derive propositions from axioms. When working on something in geometry, it is not efficient to start from the axioms every time. If you remember a few geometric things, you can always move on to other things; however, it would be more efficient to do the work in the opposite direction. Deciding which axioms are best may not be the most efficient way to solve the problem either. In physics, the Babylonian method is needed, not the Euclidean or Ancient Greek method. I would like to explain why this is so. So I can put them back together every day even if the memory weakens and some parts fall out. " It is not always efficient to derive propositions from axioms. When working on something in geometry, it is not efficient to start from the axioms every time. If you remember a few geometric things, you can always move on to other things; however, it would be more efficient to do the work in the opposite direction. Deciding which axioms are best may not be the most efficient way to solve the problem either. In physics, the Babylonian method is needed, not the Euclidean or Ancient Greek method. I would like to explain why this is so. When working on something in geometry, it is not efficient to start from the axioms every time. If you remember a few geometric things, you can always move on to other things; however, it would be more efficient to do the work in the opposite direction. Deciding which axioms are best may not be the most efficient way to solve the problem either. In physics, the Babylonian method is needed, not the Euclidean or Ancient Greek method. I would like to explain why this is so. When working on something in geometry, it is not efficient to start from the axioms every time. If you remember a few geometric things, you can always move on to other things; however, it would be more efficient to do the work in the opposite direction. Deciding which axioms are best may not be the most efficient way either, to solve the problem. In physics, the Babylonian method is needed, not the Euclidean or Ancient Greek method. I would like to explain why this is so.

The problem with the Euclid method is that the axioms are more interesting or important. But in the case of gravity we ask, for example: is it a more important, fundamental, or good axiom

that the force should be directed toward the sun, or that equal areas should be sampled in equal times? From one point of view, the statement about force is better. If I tell you what the forces are, I can study many-particle systems whose orbits are not elliptical; because
The expression force gives information about the mutual attraction. In this case, the theorem about equal areas fails. So I think that the power theorem should be taken as an axiom instead of the field. On the other hand, in the case of a many-particle system, the principle of equal areas can be generalized to another theorem. Although the meaning of this theorem is somewhat muddled and not as "elegant" as the equal areas theorem, it is clear that it derives from it. Let's take a many-particle system, for example a system consisting of Jupiter, Saturn, the Sun and many stars, all interacting with each other, and view their projections on a plane from a distance (Figure 12). The particles all move in different directions.

Figure 12

Let's pick a point and calculate how much area the radius from that point to each particle covers. Heavier masses are considered more effective in this calculation; That is, if the weight of one particle is twice that of another, its area is also assumed to be twice that of the other.

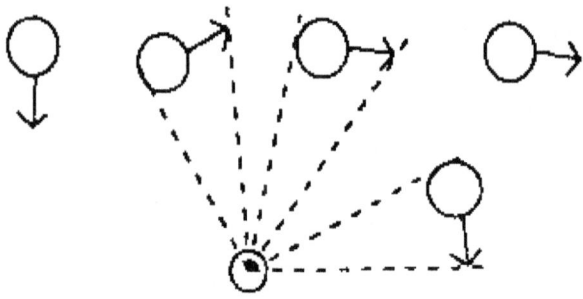

Şekil 12

So if we calculate each area scanned in proportion to the mass scanned and then add them all together, we see that the overall result does not change over time. We call this sum the angular momentum, and this rule the law of conservation of angular momentum.

Preservation means immutability. One of the consequences of this can be explained as follows: Many

Suppose the stars fall towards each other and form a nebula or galaxy. At first they are very far away, on a long radius stretching far from the center; Because they move slowly, they form a very small area. The closer they get to each other, the shorter the distance from the center, the smaller the radius when they go in, and they have to move much faster to scan the same area every second. Therefore, when stars move inward, they will fall faster. With that, we have roughly explained the nature of the shape of a spiral nebula. Likewise, we can understand how a skater spins around himself. It starts rotating slowly with one leg out and rotates faster as you pull the leg in. With its leg outstretched, it scans a specific area every second; The closer the leg is to the inside, the faster it has to rotate to create the same space. However, I am not providing a proof for the skater with this explanation. Because the skater uses muscle power, gravity is a different force; but the rule still applies to the skater.

We have a problem now. Often we can derive from some part of physics a principle that is much more fundamental than the law it came from, such as the law of gravitation. In mathematics this does not happen; Theorems don't appear where they shouldn't be. In other words, if we say that the axiom of physics is the equal-areas principle of the law of gravitation, we can derive from it the principle of conservation of angular momentum; however, this only applies to the gravitational force. However, we know experimentally that the torque is more extensive. Newton proposed other axioms from which a more general law of conservation of angular momentum could be derived. But the laws he received were wrong. There is no force in them, there is a lot of nonsense, particles do not have orbits, etc. On the other hand, the principle of field and angular momentum conservation is similar in this system and also applies to intraatomic motions in quantum mechanics. We have far-reaching policies that encompass a variety of laws. This If we take very seriously the laws we derive from principles, and assume that one is valid only because the other is true, we fail to understand the connections between the different branches of physics. One day when the physics is complete and all the laws are known we will be able to start with some axioms. No doubt someone will succeed in finding a special way to extract everything. If we don't know all the laws, we can only make guesses using some theorems for cases not covered by the proof. In order to understand physics, it is necessary to know and evaluate various statements and the relationships between them. Because the laws of physics often go beyond the realm of inference. This only loses its meaning when all the laws are known.

There is something else that is interesting and very strange about the relationship between physics and mathematics. You can come to the same conclusion by starting from different seemingly different places and doing some mathematical reasoning. We all know that. If you have axioms, you can use some theorems instead; but the laws of physics are so precisely constructed that their various expressions, which are equivalent to each other, have qualitatively different properties, which makes them very interesting. To explain this I will express my law of gravitation in three exactly equivalent but very different ways.

The first statement is that according to the equation I gave earlier, there is a force between bodies.

$$F = G \frac{mm'}{r^2}.$$

When any object sees this force, it gains a certain amount of acceleration or changes its motion every second. This is the regular expression of the law; I call it Newton's law. This statement of the law gives us
says that the force depends on something at a finite distance; contains a property that we call unlocal. The force on an object depends on how far away something else is at a finite distance from it.

We may not like the idea of a long-distance effect. How can this object know what is going on there? So let's express the laws in a very different way, called the "field method". It's hard to explain, but I still want to give you an idea of what it looks like because it says something very different. Every point in space has a number (I know it's a number, not a mechanism. That's the problem with physics: it has to be mathematical). As you go from one place to another, the numbers change. The force on an object at a point in space acts in the direction where the numbers are changing the fastest (we call it potential; the force acts in the direction where the potential is changing). Also, the force is proportional to the rate of change of potential during motion. This is only part of the statement and not sufficient. I also need to tell you how to spot potential changes. I would say that the potential changes inversely with distance from an object. But that would be a throwback to the idea of remote influencing. We can rephrase the law by saying that we don't need to know anything that's going on anywhere except a small ball. If you want to know what the potential is at the center of the ball, just tell me, how high the potential is at its surface, no matter how small the ball is. No need to look outside, just tell me the mass of the nearby potential and the ball. The rule is: Potential at the center of the sphere = mean potential at its surface - constant G (constant in the other equation) / twice the radius of the sphere we call a X the mass inside the sphere (if the sphere is small enough). I would say that the potential changes inversely with distance from an object. But that would be a throwback to the idea of remote influencing. We can rephrase the law by saying that we don't need to know anything that's going on anywhere except a small ball. If you want to know what the potential is at the center of the ball, just tell me, how high the potential is at its surface, no matter how small the ball is. No need to look outside, just tell me the mass of the nearby potential and the ball. The rule is: Potential at the center of the sphere = mean potential at its surface - constant G (constant in the other equation) / twice the radius of the sphere we call a X the mass inside the sphere (if the sphere is small enough). I would say that the potential changes inversely with distance from an object. But that would be a throwback to the idea of remote influencing. We can rephrase the law by saying that we don't need to know anything that's going on anywhere except a small ball. If you want to know what the potential is at the center of the ball, just tell me, how high the potential is at its surface, no matter how small the ball is. No need to look outside, just tell me the mass of the nearby potential and the ball. The rule is: Potential at the center of the sphere = mean potential at its surface - constant G (constant in the other equation) / twice the radius of the sphere we call a X the mass inside the sphere (if the sphere is small enough). There is a return to the idea of action at a distance. We can rephrase the law by saying that we don't need to know anything that's going on anywhere except a small ball. If you want to know what the potential is at the center of the ball, just tell me, how high the potential is at its surface, no matter how small the ball is. No need to look outside, just tell me the mass of the nearby potential and the ball. The rule is: Potential at the center of the sphere = mean potential at its surface - constant G (constant in the other equation) / twice the radius of the sphere we call a X the mass inside the sphere (if the sphere is small enough). There is a return to the idea of action at a distance. We can rephrase the law by saying that we don't need to know anything that's going on anywhere except a small ball. If you want to know what the potential is at the center of the ball, just tell me, how high the potential is at its surface, no matter how small the ball is. No need to look outside, just tell me the mass of the nearby potential and the ball. The rule is: Potential at the center of the sphere = mean potential at its surface - constant G (constant in the other equation) / twice the radius of the sphere we call a X the mass inside the sphere (if the sphere is small enough). If you want to know what is the potential at the center of the ball just tell me what is the potential at its surface no matter how small the ball is. No need to look outside, just tell me the mass of the nearby potential and the ball. The rule is: potential at the center of the sphere = mean potential at its surface - constant G (constant in the other equation) / twice the radius of the sphere we call a X the mass inside the sphere (if the sphere is small enough). If you want to know what the potential is in the middle of the ball, just tell me what the potential is at its surface no matter how small the ball is. No need to look outside, just tell me the mass of the nearby potential and the ball. The rule is: potential at the center of the sphere = mean potential at its surface - constant G (constant in the other equation) / twice the radius of the sphere we call a X the mass inside the sphere (if the sphere is small enough).

You see that this law differs from the other; because it shows what is happening at a point in relation to what is happening in the region very close to that point. Newton's law, on the other hand, states what is happening at one point in time in relation to what is happening at another point in time. It expresses what is happening from one moment to the next in relation to what is happening from one place to the next. The second statement is local in both time and space; because it only depends on those in the immediate vicinity. Mathematically, however, the expressions are completely identical.

There is a completely different way of putting it, in terms of the philosophical and qualitative ideas it contains. If you don't like the remote effect idea I'm talking about, you can do without it. Now I'm going to give you a statement that, from a philosophical point of view, is the complete opposite of that. There is no discussion at all of how the objects get from one place to another; all included in the following general statement. If you have multiple particles and you want to know how one of them travels from one place to another, you do it by finding a path between those two points in a given time (Figure 13).

Figure 13

Suppose the particle wants to get from X to Y in one hour. They want to know which way to go about it.

Şekil 13

What you're going to do is come up with several curves and calculate a specific quantity for each one (I don't want to talk about what that quantity is, but for those who know the terms I'll tell you that for each curve it's this is size is the mean of the difference between kinetic and potential energy). If you calculate this number for one route and then another, you will find a different number for each route. There is a path that gives the smallest number possible, and that is the path an object actually takes in nature! Now we're talking real motion, the ellipse, the total orbit. Causality is no longer in question as the particle feels gravity and moves under its influence. Instead we can say:

You will see an example of nature expressing itself in various incredible ways. If causality is to be necessary in nature, you can apply Newton's law; or when nature is said to be narrated with the minimal principle, speak after this last form; or if it is emphasized that nature should be a local area, you can do that too. The question is: which one is correct? When these different approaches are not mathematically equivalent, when the results on some differ from those on others, we need only conduct experiments to find out which nature actually chose. Arguments can be made that philosophically favor one over the other;

However, we have learned from long experience that all philosophical insights into what nature will do fail. It is necessary to explore all possibilities and try different ways. In this particular example in question, the theories are all exactly equivalent. All three forms of expression; Mathematically, Newton's law, the local field method and the minimum principle give exactly the same result. What do we do now? You can read in all the books that we cannot make a scientific decision for one or the other. This is true. All three are scientifically related. Mathematically, Newton's law, the local field method and the minimum principle give exactly the same result. What do we do now? You can read in all the books that we cannot make a scientific decision for one or the other. This is true. All three are scientifically related. Mathematically, Newton's law, the local field method and the minimum principle give exactly the same result. What do we do now? You can read in all the books that we cannot make a scientific decision for one or the other. This is true. All three are scientifically related.

is equivalent. It is impossible to make a choice as the results are the same and there is no empirical way to distinguish. Psychologically, however, they are very different in two respects. First, for philosophical reasons, you either love them or you don't. The only way to prevent this disease is education. Second, the terms are psychologically distinct; because there can be no equivalence between statements aimed at finding new laws.

As long as physics is not complete and we are trying to understand other laws, different expressions of laws can give us clues about what might happen in different situations. To help us appreciate what form laws will take in more general situations, the statements are not equivalent. An example; Einstein realized that electrical signals cannot travel faster than the speed of light. He predicted that this was a general principle (it's like taking angular momentum and generalizing it from one proven event to all events in the universe) and that if it holds true for everything, it would also hold for gravity . If signals couldn't travel faster than light, it wouldn't be enough think about power for a moment. Explaining physics in Newton's terms, when considered in the context of Einstein's generalization of the law of gravitation, it is very inadequate and its operations become very complex. The field method expression is neat and simple; so with the principle of the minimum. We haven't decided between these two yet.

In fact, both turns out to be not quite true for quantum mechanics, as I said before. However, it turns out that the existence of a minimal principle is the result of very small particles moving according to quantum mechanics. As far as we can tell, the best law is a combination of the minimum principle and local laws. Well, both the local features of the laws of physics and those
we think they should include the principle of the minimum; but we are not sure yet. If you have a result that is only partially true and you see that something will go wrong, write it by choosing the right axioms, maybe only one axiom will turn out to be invalid and the others will remain. In this case, you only need to change one small thing. But if you write it with other axioms, the whole structure can collapse; because everything is based on this one axiom, which is false. We cannot predict the best choice unless we have an intuitive inclination to learn about new situations. We must not fail to look at something from all possible angles; This means,

One of the most surprising features of nature is the variety of possible systems of interpretation. It is understood that this is due to the nature of the laws, their special and delicate nature. For example, the inverse square of the law makes it local; This would not be possible with an inverted cube. On the other hand, what allows us to write the laws using the principle of the minimum is that in the equation force is associated with change in velocity. If the force were not proportional to velocity but, for example, to the rate of change of position, we could not write it like this, as the principle of minimum. If you change the laws too much, you'll find that you can write them in fewer different forms. That always seemed puzzling to me and I couldn't understand why the true laws of physics can be written in so many different ways. It's like they can manage to go through different doors at the same time.

I would like to say a few more general things about the relationship between mathematics and physics. Mathematicians only deal with the structure of thought. They don't care what they talk about. The things they talk about; or, as they admit, they don't even have to investigate whether what they say is true.

they don't see. I will explain this. They state the axioms: this and that is so, this and that is so. Then? One can argue without knowing the meaning of the words this and that. When statements about axioms are carefully formulated and sufficiently complete, the thinker does not need to know the meaning of words to draw new conclusions in the same language. If I use the word triangle in one of the axioms, the result will be a statement about the triangle; whereas the person arguing may not know what a triangle is. I can analyze its logic and say, "The triangle is this three-sided thing, something like that"; so I can learn new results. In other words, if you have axioms about the real world, the mathematician gives you lines of reasoning, that you can use. The physicist gives meaning to all his propositions. This is an important point that those who approach physics through mathematics fail to realize. Physics is neither mathematics nor mathematics physics; they just help each other. In physics, you need to know how words relate to the real world. In the end you have to translate your result into everyday language, into the world, onto the copper or glass pieces you are experimenting with. This is the only way to determine if the results are correct. This is a problem that has nothing to do with mathematics. This is the only way to determine if the results are correct. This is a problem that has nothing to do with mathematics. This is the only way to determine if the results are correct. This is a problem that has nothing to do with mathematics.

The mathematical considerations developed so far are undoubtedly very important and useful for physicists. On the other hand, the reasoning of physicists is sometimes useful to mathematicians.

Mathematicians want to make their reasoning as general as possible. If I say to them, "I want to talk about normal three-dimensional space," they say, "There are these theorems for n-dimensional space." "I only want for 3", "so replace n with 3! Finally, many of the complex theorems become much simpler when applied to special cases. Any physicist
time deals with special cases; It doesn't care about general situations at all. He's talking about "something," not something abstract. He wants to examine the law of attraction three-dimensionally; not what the force will be in n dimensions. So some kind of reduction is necessary, because the mathematician prepared them for larger problems. That's a very useful thing. At some point the poor physicist has to say, "Excuse me if he wants to tell me about the four dimensions..."

If you know what you are talking about ie some symbols have power, some have mass, inertia etc. You can use your common sense if you know what it stands for. You've seen a lot of different things and you pretty much know how it's going to play out. But the poor mathematician turns it into equations; Since symbols mean nothing, there is no guidance other than precise mathematical rigor to guide his discussion. The physicist who knows more or less what the answer will be moves faster, partly by prediction. Although very precise mathematical accuracy is not very useful to the physicist, one should not criticize the mathematician in this regard. Just because something will be useful to physics doesn't mean the mathematician has to do it that way. He minds his own business. If you want something else, you can do it yourself.

Another question is whether one should use preconceived approaches and philosophical principles when trying to find a new law; E.g. "I didn't like the minimal principle" or "I liked the minimal principle", "I didn't like the remote effect" or "I really liked the remote effect". How do models help? It is interesting that models are mostly useful. Many physics teachers also try to teach ways to use models to get a good idea of how everything works. But it also shows that the greatest inventions are created independently of the model and the models are useless. Maxwell's The discovery of electrodynamics was first made by imagining that many wheels and pulleys were moving through space. Throw aside all roles and everything else in space, the theory still holds true. director[12]found the laws of relativistic quantum mechanics by simply guessing the equation. The method of estimating the equation seems to be an efficient way to estimate new laws. This shows us once again that mathematics is a way of deeply expressing nature, and that a preconceived approach to philosophical principles is not an efficient way.

There is always something that makes me think. According to the law as we understand it today, it takes an infinite number of logical functions of a computer to know what is going on in even the smallest part of the universe and in the shortest possible time. How can all this happen in such a small space? Why does it take infinite amounts of logic to understand what's going on in a small piece of space-time? I often hypothesize that physics doesn't need a mathematical expression, that eventually how it works will become clear, and that although it looks like a complicated chessboard, the laws will turn out to be simple. I know these predictions are of the same nature as other people's "like" or "dislike" statements. I think it's not good to be too biased about these things.

In conclusion, I feel compelled to use Jean's words: "The great architect must have been a mathematician". For those who are not versed in mathematics, it is very difficult to really feel the beauty and the deepest beauty of nature. CP Snow spoke of two cultures. I think these two cultures divide people into those who are experienced enough in understanding mathematics to really appreciate nature and those who aren't.

Unfortunately, physics has to be mathematics; Mathematics is also difficult for some people. That being said - is it true? I do not know. A king trying to learn geometry from Euclid complained about the difficulty. Euclid replied, "There is no 'royal road' to geometry." There really is no royal road. Physicists cannot turn to any other language. If you want to learn about nature in order to understand it, you have to understand the language it speaks. He conveys his messages in only one way. In order for us to listen to him, we must not be so humble as to ask him to change first.

No intellectual explanation can convey to deaf ears what an experience music is. Likewise, all the intellectual explanations that can be made in the world cannot make people from the "other culture" understand nature. Philosophers may try to teach nature in terms of quality; I try to explain it too. But I can not; because it is not possible. Some people imagine that the center of the universe is man, perhaps because their horizons are so limited.

Important Conservation Principles

As you learn the laws of physics, you will find that there are many complex and detailed laws such as: B. the laws of gravity, electricity and magnetism and nuclear interaction. There are some great general principles that override and govern all of these elaborate laws. Examples of these principles are conservation principles, certain symmetry properties, the general form of quantum mechanical principles and, for better or worse, in the last lecture we discussed.
we can count the fact that all the laws we get are mathematical. In my talk today I will talk about conservation principles.

Physicists use normal words in a different way. For them, the law of conservation means: at any moment there is a number that you calculate. Then, after nature has changed by the masses, if you calculate this quantity again, you will find the previous number; the number does not change. An example of this is the law of conservation of energy. There's a number that we calculate according to a certain rule, and no matter what, the number is always the same.

You might find that something like this could be useful. Let's imagine that physics or nature is like a chess game with millions of pieces. We want to know according to which laws the stones act. It is difficult to see and understand how fast the great gods play this chess game. But we can still see some rules, there are also some rules we can find without looking at every move. For example, there is only one elephant, a white elephant, on the board. Since the elephant always moves diagonally, the color of the square it is in never changes. If we turn our heads away for a while while the gods are playing, when we look back at the board later, we can expect the white elephant to be somewhere else, but on a square of the same color. It's something of a conservation principle. We don't have to search all the time to know at least some things about the game.

The chess example is perhaps not quite correct for this law. If we look away for a long time, maybe the bishop lost, the pawn got promoted to queen, and the god maybe decided it would be better to have a bishop instead of a queen where that pawn is on the black square. Unfortunately, it may turn out later that some laws are not entirely correct today. But I will tell you the laws as we know them today.

I told you that we use everyday words in a technical sense. One of the words in the title of today's talk (Great Conservation Tics) is the word big. This is not a technical term; it was only placed there to make the headline more effective. I could have called it "Laws of Conservation" instead. There are several not fully working conservation laws that are approximately correct; However, in some cases they are useful. We can also call them "minor" conservation laws. I'll talk about some of these later. The basic laws that I will discuss next are spot on as we know them today.

I will begin my discussion with the conservation of electric charge, which is the easiest to understand. There is a number that expresses the total electric charge in the world, and that number doesn't change no matter what. If you lose it in one place, you will find it somewhere else. Conservation applies to the total electric charge. This is experimental Faraday[13] was discovered by The experiment consisted in determining the electric charge of the sphere by penetrating a metal sphere with a very sensitive galvanometer on its outer surface, as a small charge would have a large effect. Faraday placed various strange electrical devices inside the sphere. He created a charge by rubbing glass rods on the cat's skin and placed large electrostatic machines inside the sphere, reminiscent of laboratories in horror movies. However, during all these experiments no electric charge was formed on the surface; there was no net electric charge. The glass rod could have been positive by rubbing it against the cat's skin; then the post would also be negative and the total charge would always be zero. If there was a charge inside the sphere, would this affect the outside of the galvanometer. The total load was therefore retained.

That's easy to understand; A very simple non-mathematical model would be illustrative. That the world is made up of two types of particles, electrons and protons - it used to be thought that everything was so simple - and to tell them apart, electrons have to be

Suppose the protons carry a positive charge. We can take an object and give it more electrons, or we can take some of it. Assuming the electrons are permanently or not destroyed—this is a simple statement that isn't even mathematical—the difference between the total number of protons and the total number of electrons doesn't change. In this model, neither the number of protons nor the number of electrons changes. We are interested in the electric charge! The contribution of the protons is positive and that of the electrons is negative. Since there is no self-generation or self-destruction for these objects, the overall charge is conserved. In this talk, I want to list some properties whose sets do not change. I'll look at the load first (Figure 14). The table that Professor Feynman completes with additions during the lecture

Figure 14

I write "yes" in front of the question whether the load is protected or not. The theoretical explanation is very simple; However, it was later discovered that electrons and protons are not permanent. For example, a particle called a neutron can split to give an electron and a proton - another thing I'll talk about later. However, the neutron is electrically neutral. Therefore, the number of electrons and protons is not constant as they can be created from neutrons, but the charge is still the same. As a result, we started with a sapphire charge, and then again we had a plus one and a minus charge with zero sums.

Another similar phenomenon is observed with another positively charged particle, such as a proton. This particle, which is a kind of replica of an electron and resembles an electron in many ways, is called a positron. However, unlike an electron, its charge has an opposite sign, and more importantly, when an electron and a positron meet, they annihilate each other, resulting in just one light. For this reason, the positron is called an antiparticle. This means that electrons are not even invariant by themselves. The electron plus positron becomes light. "Light" is actually an invisible gamma ray. But for a physicist it's the same, only the wavelength is different. So we see that a particle and its antiparticle disappear. Light has no electric charge; but since we took a positive and a negative charge, the overall charge didn't change. For these reasons, the theory of conservation of charge is still non-mathematical, albeit somewhat complex. You add up the number of positrons and protons you have and subtract the number of electrons from that. There are other particles you need to check. for example counter protons making a negative contribution, pi plus mesons making a positive contribution. Actually all elementary particles in nature You add up the number of positrons and protons you have and subtract the number of electrons from that. There are other particles you need to check. for example counter protons making a negative contribution, pi plus mesons making a positive contribution. Actually all elementary particles in nature You add up the number of positrons and protons you have and subtract the number of electrons from that. There are other particles you need to check. for example counter protons making a negative contribution, pi plus mesons making a positive contribution. Actually all elementary particles in nature has a charge. All we have to do is add their numbers. Whatever happens in any reaction, the total charge on one side must be balanced with that on the other side.

    This is an aspect of conservation of electricity. Now we face an interesting question. Suffice it to say that the charge is conserved, do we need to add anything else? If the charge is conserved, if it is a real moving particle, it has a completely different property. The total charge in a box can stay the same in two different ways. Maybe the cargo in the box is moving from place to place. Another possibility is that the charge in one place disappears and at the same time appears in another place related to that moment, so the total charge does not change. In terms of conservation, this second possibility differs from the first in that; If the charge disappears in one place and reappears in another, something must be moving in the gap. This second form of charge conservation is called local charge conservation and contains much more detail than simply saying that the overall charge does not change. If it's true that the payload is locally protected, let's improve our law. This is indeed true. From time to time I have tried to show you ways of reasoning, ways of connecting one thought to another. Now I want to tell you about another argument, caused mainly by Einstein, that if something is conserved - I apply it to the load in this example - it must be locally conserved. This reasoning rests on the fact that the question of which of two people passing by spaceships is moving and which is stationary cannot be determined experimentally. This,

Suppose there are two spaceships A and B (Figure 15). I think A is overtaking B.

I approve.

Figure 15

Note that this is just one view, you will see the same natural phenomenon if you look the other way. Now suppose that the person standing still wants to discuss whether they saw a charge disappear at one end of the spaceship and appear at the other end at the same time. Also, let's assume that the man made sure to sit right in the middle of the ship to make sure it happened at the same time. If he sat in front of the ship he would have seen one before the other as it took some time for the light to come on. There is another man on the other ship making the same observations. Now lightning struck and a charge formed at the x-point and at the same time the charge at the y-point disappeared, which is the other end of the ship. Note that I use the term "simultaneously" to conform to the principle of conservation of charge. If we lose an electron in one place, we gain an electron in another; but nothing goes between these two points. Suppose there's a glow where the charge disappears and reappears so we can see what's going on. B will say that the two glows are simultaneous; because it is in the middle of the ship and the light that appears in x and disappears in y from the flash there is a glow where the charge disappears and reappears so we

can see what is going on. B will say that the two
glows are simultaneous; because it is in the
middle of the ship and the light that appears in x
and disappears in y from the flash there is a
glow where the charge disappears and reappears
so we can see what is going on. B will say that
the two glows are simultaneous; because it is in
the middle of the ship and the light that appears
in x and disappears in y from the flash
He knows that he will reach him at the same
time as the light. B will say: "Yes, when one
disappeared, the other appeared. "And what
about our friend on the other ship? "No, you are
wrong, my friend. I saw the glow of x before y,"
he will say. The reason for this can be explained
as follows; Because it is moving toward x itself,
the light from x has traveled a shorter distance
than that from y; because B is moving away
from y. He might also say, "No, I think x flashed
first, then y disappeared. Therefore, a charge
was generated in the short time that elapsed
between the flashing of x and the disappearance
of y. This is not charge conservation; is against
the law. The first man says: "Yes, but you
move", the other says: "How do you know that?
I think you're moving", etc. If we are unable to
establish by any experiment that it makes a
difference in the laws of physics whether we
move or not, then the conservation of charge

must be local. If it weren't so, it would only apply to a few people, namely the man who stands absolutely still. Since this is impossible according to Einstein's relativity principle, non-local conservation of charge is also not possible. The locality of the conservation of charge corresponds to the theory of relativity and applies to all conservation laws. If something is not pasted, you can understand that the same principle applies. The locality of the conservation of charge corresponds to the theory of relativity and applies to all conservation laws. If something is not pasted, you can understand that the same principle applies. The locality of the conservation of charge corresponds to the theory of relativity and applies to all conservation laws. If something is not pasted, you can understand that the same principle applies.

There is another very interesting thing about electric charge; something unrelated to conservation law, regardless, which we haven't found any real explanation for yet. The load is always in units. A charged particle has a charge or two; or minus one or minus two fees. Although this has nothing to do with preserving the payload, to get back to our table I need to write in the table that the protected thing occurs in units. Being in units makes the theory of conservation of charge easy to understand. We can count that when we go from one place to another. Finally, from a technical point of view, it is possible to easily determine the total charge of something electrically. Because it's a very important one It has a function. Charge is the source of electric and magnetic fields. Charge is a measure of a body's interaction with electricity, with an electric field. Another thing to add to our list is that charge is a field source, electricity is charge dependent. So the particular amount that is preserved here has two interesting aspects, but not directly related to the preservation. The first is that it comes in units and the second is that it is the source of a field.

There are many conservation laws. I will give some examples of laws of the same kind as conservation of charge, where countability alone will suffice. There is a conservation law called the conservation of baryons. A neutron can become a proton. If we denote the charge in each of these units as a unit or baryon, the baryon number has not changed. The neutron carries or represents a baryon charge unit; a proton represents a baryon - all we do is count and use fancy words. When the reaction I'm

(kolay) $P + P \rightarrow \lambda + P + K^+$

talking about takes place, that is, when a neutron becomes a proton, dissociates into an electron and an antineutrino, the total number of baryons doesn't change. There are other reactions in nature. A proton plus a proton can create a variety of strange objects; For example a lambda, a proton and a K+ Lambda and K+ denote the odd particles.

In this reaction we see that we put two baryons and only one remains; So either lambda or K+ has a baryon. After that, if we examine lambda, we see that it very slowly decays into a proton and a pi, and finally pi decays into electrons and other things.

$$(yavaş) \quad \lambda \rightarrow P + p$$

We see here how the baryon reappears in the proton. So we think lambda has a baryon number of 1, but since there is no proton in K+, the baryon number is zero.

So we can put the load on our conservation law table (Figure 14). A similar situation exists with baryons. But according to the special rule here, the baryon number is calculated as follows: The number of protons plus the number of neutrons plus the number of lambdas minus the number of antiprotons minus the number of antineutrons, etc. This is just a counting operation. Baryons are also conserved and spawn in units; also, although no one knows why, everyone likes to think it's analog to a field source. The reason we made these tables is that we are trying to discover the laws of nuclear interaction and the tables make it easy to make predictions. If the charge is a field source and the baryon otherwise has the same properties as a charge, it must also be a field source. But unfortunately it doesn't look like that at first; It's possible it is, but we don't have enough information to be sure.

While there are a few other counting sets (e.g. lepton numbers, etc.), the basic idea is the same as with baryons. However, there is something else that is slightly different. The reactions between these strange particles in nature have their own reaction rates; Some reactions are very quick and easy, while others are very slow and difficult. I am using the words easy and difficult in a technical sense, not in reference to how the experiments are run, but in reference to how fast the reaction is going. There is a distinct difference between the two reactions mentioned above, the decomposition of the proton and the much slower decomposition of the lambda. There is also a number rule for quick and easy reactions. So lambda becomes minus one plus one and proton becomes zero.

seems to be true for reactions and false for difficult reactions. We need to add to our table (Figure 14) the conservation law called conservation of strangeness or conservation of hyperon number, which is almost correct. It is clear why we characterize this size as strangeness. It is almost true that it was preserved, it is true that it appeared in units. In trying to understand the strong interaction between the nuclear forces, the fact that something is conserved in the strong interaction made some people think it could also be a field source. However, we don't know for sure. I gave this example to show you how conservation laws can be used to predict new laws.

There are other conservation laws based on counting that are sometimes proposed. For example, chemists once thought that the number of sodium atoms does not change in any way. However, sodium atoms are not invariant. It is possible to convert an atom of one element into an atom of another element in such a way that the original element is completely destroyed. Another law that was held to be true for a while was that the total mass of an object does not change. It depends on how you define mass and whether you confuse it with energy. The law of conservation of energy, which I will discuss in a moment, also includes the law of conservation of mass. Of all the conservation laws, the law of conservation of energy is the most difficult and abstract, but also the most useful. It's harder to understand than that what I explained before because the mechanism of responsible and other conservation laws is obvious; This mechanism is the preservation of objects in one form or another. The situation is quite different when it comes to energy. Because out of old things we get new things; however, it is still just a simple counting operation.

Conservation of energy is a bit more difficult; because here too we have a number that does not change over time, but this number does not mean anything special.
does not represent. To explain this I will use an analogy that seems absurd.

I want you to think that a mother would leave her child alone in a room with 28 indestructible toy cubes. The child plays with the dice all day, and when the mother comes home, she constantly checks whether the dice are "protected" and finds that there are actually 28 pieces. The same is repeated for several days. When mom comes home one day, she counts 27 dice! But in front of the window he finds a cube that the boy threw away. The first thing to understand in conservation laws is to be careful that what you are trying to control has not gone out. If another child had come to play with some dice, the same could have gone the other way as well. Of course we have to take things like this into account when we talk about conservation laws. Suppose one day when she comes home, mom finds only 25 cubes. He suspects that the boy has hidden three cubes in a toy box. "I'll open the box," he says. The child replies, "No, you can't open the box." Anne is a very smart person; says: "When the can is empty, 16 ounces[14] I know it comes in and each cube is 3 ounces. I'll weigh the box now."

Taking all the dice into account, he applies his formula and finds the sum of 28. It goes on like this for a while. But one day the total doesn't come out right. In the meantime, he notices that the dirty water level in the sink has changed. He knows that the depth of water without a cube in it is 6 inches, and that the level rises 1/4 inch when there is a cube in the water. Adds another term to the expression:

There are 28 in total again. The more cunning the child becomes and the mother remains just as cunning, the more terms have to be added. All terms represent the number of dice. However, these are mathematically abstract calculations because the cubes are not visible.
I will now make the analogy and tell you

what is common and different between this example and conservation of energy. First of all, let's assume that we don't see any dice anyway. The term "visible dice" is not included in the total. In this case, the mother constantly says "cubes in a box", "cubes in water", etc. it will continue to calculate many terms. Here's how it differs from energy: As far as we know, there are no dice. Also, unlike dice, the numbers found for energy are not whole numbers. It may be that when the poor mother calculated one term, she found the result 6 1/8 7/8 for another term and 21 for the others. There will be 28 in total again. It is similar with energy.

What we discovered for energy is a system that consists of a set of rules. We find numbers determined by different rules for different energies. If we add up all the numbers we've found for different types of energy, the sum is always the same. However, as far as we know, there are no real units, little marbles. The process is purely abstract and mathematical. Every time we do arithmetic, there's always a number that's about to come out. I can't explain it any better.

This energy is the dice in the box, the dice in the water, etc. It manifests itself in different forms. The energy that we call kinetic energy as a result of motion, the energy that arises as a

result of gravitational interaction (we call it gravitational potential energy), thermal energy, electrical energy, spring, etc. elastic energy, chemical energy,
There is nuclear energy, and there is energy that arises from the mere existence of the particle and depends on its mass. The latter is the famous equation $E = mc^2$ contributed by Einstein, as you no doubt know.

    I have spoken of many energies; However, I would like to add that we are not totally ignorant on this subject, we understand the relationship of some to others. For example, what we call thermal energy consists largely of the kinetic energy of particles in a body. Elastic energy and chemical energy have the same origin: forces between atoms. When atoms take place in a new order, some energy change occurs. This means that if this quantity changes, another quantity must also change. For example, when we burn something, the chemical energy changes; Heat is created where there was no heat before. Because everything has to add up to the right sum. Elastic energy and chemical energy both come from interatomic interaction. Now one of these interactions is electrical energy, we have understood that it is a combination of two different energies, the other of which is kinetic energy. But this is a quantum mechanical formula. Light energy is nothing other than electrical energy, because light is now interpreted as electrical and magnetic waves. Nuclear energy cannot be expressed by others; I can't say much at the moment that it's the result of nuclear forces. What I am talking about is not the energy released. There is a certain amount of energy in the uranium nucleus. When uranium fissions, the energy remaining in the core changes; but the total energy in the world does not change. Meanwhile, the resulting heat etc. It also provides balance. because light is now interpreted as electric and magnetic waves. Nuclear energy cannot be expressed by others; I can't say much at the moment that it's the result of nuclear forces. What I am talking about is not the energy released. There is a certain amount of energy in the uranium nucleus. When uranium fissions, the energy remaining in the core changes; but the total energy in the world does not change. Meanwhile, the resulting heat etc. It also provides balance. because light is now interpreted as electric and magnetic waves. Nuclear energy cannot be expressed by others; I can't say much at the moment that it's the result of nuclear forces. What I am talking about is not the energy released. There is a certain amount of energy in the uranium nucleus. When uranium fissions, the energy remaining in the core changes; but the total energy in the world does not change. Meanwhile, the resulting heat etc. It also provides balance.

The law of conservation of energy is very useful in many technical fields. I will try to explain with some simple examples how knowing this law and the energy calculation formulas will help us to understand other laws. In other words, others
many laws are not independent; they are just a secret expression of conservation of energy. The simplest of these is the law of leverage. (Figure 16).

Figure 16

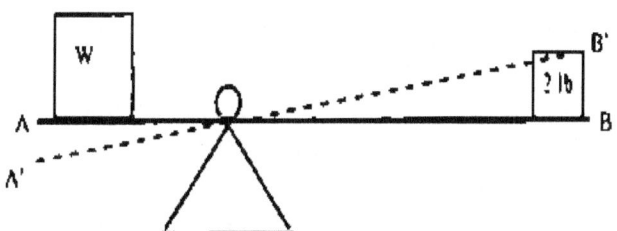

We have a lever at a fulcrum. One arm is 1 foot long and the other 4 feet. First I need to state the law of gravitational energy. If you have multiple weights and multiply the weight of each by the height from the ground and add them all together, you get the gravitational energy. 2 pounds in the long sleeve[fifteen] have a weight and an unknown weight on the other side - X always denotes the unknown, but let's call it W to look unusual! Our question is: How long does W have to be so that the lever can freely swing up and down in full equilibrium? If the lever swings up and down slightly, it means that even if the lever is parallel to the ground, the energy is always the same no matter the 2lb weight is, say, 1 inch off the ground. If the energy is always the same, it doesn't matter what position the lever is in and the weights won't fall. How low does the W go when the 2 lb weight goes up an inch? As we can see in the picture (Figure 16), the OA is 1 foot. and the OB length is 4 feet. If yes, BB' is 1 inch and AA' 1/4 inch. Now let's apply the law of gravitational energy. All heights were zero before anything happened; so the total energy was also zero. 2 lb to find the gravitational energy after the start of motion. Multiply the weight by 1 inch height, add the unknown weight W times -1/4 inch height. This sum should give us the previous energy zero. So 2 - W/4 = O, so W must be 8. Understanding this simple law we all know as the Law of Leverage add the unknown weight W times -1/4 inch height. This sum should give us the previous energy zero. So 2 - W/4 = O, so W must be 8. Understanding this simple law we all know as the Law of Leverage add the unknown weight W times -1/4 inch height. This sum should give us the previous energy zero. So 2 - W/4 = O, so W must be 8. Understanding this simple law we all know as the Law of Leverage this is the way. It is interesting, however, that not only these, but hundreds of other physical laws are related to different forms of energy. I gave this example to show how useful it is.

However, this only has one downside; In practice, it does not give accurate results due to friction at the pivot point. When I have something moving, for example a ball rolling on a horizontal plane, it stops after a while due to friction. Where does the kinetic energy of the ball go? Answer: The kinetic energy of the ball is transferred to the vibrational energy of the atoms in the ball and on the ground. The world we see is, on a grand scale, like a big ball of polish. On a small scale, it is very complex: billions of tiny atoms of different shapes, on closer inspection it resembles a rough stone; because it consists of small balls. The ground conditions are the same; lumpy balls... If we roll this huge stone over the enlarged surface, you see, that after all that pushing and shoving, the atoms still vibrate a little; that is, there is a vibrational motion or thermal energy that remains on the ground. Even if at first glance it seems that the conservation law does not apply, energy seems to have a tendency to hide from us. A thermometer and some other tools are needed to make sure it's still there. No matter how complex the process, even if we don't know the details clearly, we see that the energy is fully preserved. A thermometer and some other tools are needed to make sure it's still there. No matter how complex the process, even if we don't know the details clearly, we see that the energy is fully preserved. A thermometer and some other tools are needed to make sure it's still there. No matter how complex the process, even if we don't know the details clearly, we see that the energy is fully preserved.

The law of conservation of energy was first explained not by a physicist but by a medical doctor. For this he used mice. You can see how much heat is generated when food is burned. If you feed rats a certain amount of food, the food will turn into carbon dioxide under the influence of oxygen, just like when it is burned. If you measure the energy, the energy in both cases, you will see that living beings do the same with inanimate objects. The law of conservation of energy also applies to other phenomena.
It applies to life as it does to life. I'd also like to add that it's very interesting that every law we know that applies to "inanimate" things turns out to be true, even when tested for this great phenomenon called life. In the context of the

laws of physics, there is still no insight that requires what is much more complex in living beings to be different from what happens in non-living beings.

The amount of energy in food is how much heat, mechanical work, etc. It is measured by the "calorie" which determines how much I will bring out. Calories indicate how much heat energy is in food. Physicists sometimes seem so superior and self-confident that others want to make up for their weaknesses somewhere. Now let me tell you something that will help you catch them. Physicists should be ashamed of taking energy and measuring it in so many ways and with so many different names! Measuring energy in calories, ergs, electron volts, kilogram meters, BTUs, horsepower hours, kilowatt hours is nonsense. That means money sterling, dollars etc. It reminds me of measuring. However, there is a difference; The ratios of currencies to each other can change for economic reasons, and the ratios of all these different units of energy to each other are absolutely constant. To take an example, they are like pounds sterling and shillings; A pound is always twenty shillings. But physicists use an incredible ratio of 1.6183178 for 1 pound instead of 20 as the ratio. You would think that at least modern high-level theoretical physicists could use a common unit;

but you will see articles using Kelvin degrees for energy measurements, megacycles and, in the most recent fashion, inverse Fermis. For those wanting to prove that physicists are human too, the proof is the nausea of using such a variety of units to measure energy. but you will see articles using Kelvin degrees for energy measurements, megacycles and, in the most recent fashion, inverse Fermis. For those wanting to prove that physicists are human too, the proof is the nausea of using such a variety of units to measure energy. but you will see articles using Kelvin degrees for energy measurements, megacycles and, in the most recent fashion, inverse Fermis. For those wanting to prove that physicists are human too, the proof is the nausea of using such a variety of units to measure energy. to use such a variety of units to measure energy. but you will see articles using Kelvin degrees for energy measurements, megacycles and, in the most recent fashion, inverse Fermis. For those wanting to prove that physicists are human too, the proof is the nausea of using such a variety of units to measure energy. to use such a variety of units to measure energy. but you will see articles using Kelvin degrees for energy measurements, megacycles and, in the most recent fashion, inverse Fermis. For those wanting to prove that physicists are human too, the proof is the nausea of using such a variety of units to measure energy.

There are some natural phenomena that pose interesting energy problems. A so-called quasar was recently discovered. They are very, very far away and are emitting tremendous energy in the form of light and radio waves. Where does this energy come from? If the law of conservation of energy is true, then the state of the quasar after giving off this amount of energy should be different than its previous state. The question is; Is this due to gravitational energy? Has this object gravitationally collapsed or is it in a different gravitational state? Or is it all because of nuclear energy? Nobody knows these. They will say that maybe the law of conservation of energy is wrong. When little-studied things like quasars—quasars are so distant that astronomers can't easily see them from there—when they break basic laws, it's very unlikely that they're due to flaws in the laws; The reason is often that the details are not known.

Another interesting example of the application of the law of conservation of energy is the reaction in which a neutron decays into a proton, an electron and an antineutrino. It used to be thought that the neutron turns into a proton plus an electron. However, the energies of all particles could be measured, and the sum of a proton and an electron would not result in a neutron. In this case there were two options. Maybe the law of conservation of energy wasn't right. Indeed sometime Bohr[16]He suggested that the law of conservation of energy can only apply statistically to average values. However, it has now been established that the second possibility is correct and that the reason the energy calculation does not work is that something else has arisen, which we call the antineutrino. This counter neutrino also received energy. One could say that the only reason the antineutrino occurs is to make the law of conservation of energy true. But this particle also makes many other things true; for example conservation of momentum and other conservation laws. On the other hand, it has recently been proven that neutrinos actually exist.

This example illustrates a point. How is it possible for us to extend our laws to areas we believe in?
After confirming the conservation of energy somewhere, why are we so sure that a new situation must also obey the law of conservation of energy? Sometimes you read in the newspapers that physicists have discovered that a law they love so much is wrong. So is it wrong

to say that a law applies in an area that has not yet been studied? If you can never say that a law is true in an area we haven't examined, you know nothing. If the laws we find consist only of laws we have just finished observing, we can never make forward-looking predictions. However, the only benefit of science is to go ahead and try to make predictions. So we always take the risk. As for energy

This means the science is inaccurate. You must also be imprecise if you make a suggestion about a review that you haven't seen directly. But we also have to make statements about areas that we don't see; Otherwise all these things are useless. For example, the mass of a moving body changes due to the conservation of energy. due to the relationship between energy and mass, the kinetic energy exerts an additional mass effect; thus moving objects become heavier. Newton thought that this was not the case, that the masses remained constant. When Newton's idea turned out to be wrong, everyone kept saying that it was a terrible thing to see physicists make mistakes. why did they think would you be right? In reality, the effect is very small and only occurs as you approach the speed of light. When we spin a top, the mass is the same as without it, with a very, very small margin of error. So should it be said that "the mass doesn't change unless you spin it faster than this or that"; would it be safe then? No, because if the experiment had only been done with wood, copper, and steel tops, "Wood, copper, and steel unless the tops made move faster than this or that etc..." would have to be said. As you can see, we don't know all the conditions necessary for an experiment. We don't know if the mass of a radioactive spinner is conserved. We have to make predictions to give science any use. If we don't just want to explain the experiments we're doing, we have to propose laws that apply beyond the scope of their observation. There's nothing wrong with that, even if it keeps science from certainty. If you thought the science was too accurate, you were wrong.

Returning to our list of conservation laws (Figure 14), we can also add energy to it.

As far as we know, energy is totally conserved, it doesn't come in units. The question now is whether it is a field source. The answer is positive. Einstein realized that my gravity was created by energy. Energy and mass are equivalent; Therefore, Newton's interpretation that mass causes gravity was changed to mean that gravity is caused by energy.

There are other laws that are similar to conservation of energy in that they are made up of numbers. One of them is momentum. If you multiply all the masses in an object by their velocities and add them all together, you find the total momentum of the particles and the total amount of momentum is conserved. In the meantime it has been recognized that energy and momentum are very closely linked. That's why I put them in the same column in our table.

Another conserved quantity is angular momentum, which we considered earlier. Angular momentum is the range that moving objects scan in one second. For example, take a moving object and any center point. The growth rate of the area covered by the line connecting the center and the object is multiplied by the object's mass.

and when the numbers found in this way are added up for all objects, this is called angular momentum (Figure 17).

Figure 17

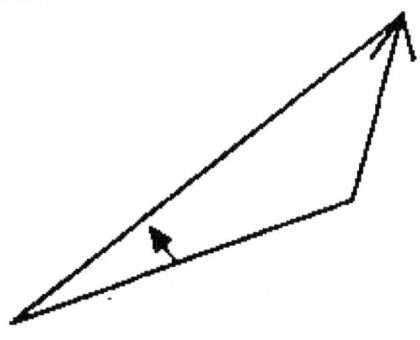

Şekil 17

This amount does not change; that is, angular momentum is conserved. At first glance one might think that angular momentum is not conserved. Like energy, it manifests in various forms. While most people think it comes in motion, I'm going to show you that it can happen in other ways. If you have a wire and you bring a magnet close by, the current through the wire and the magnetic field will increase, creating an electric current (that's how generators work). Now, instead of a wire, consider a disk with electric charges similar to the electrons in the wire (Figure 18).

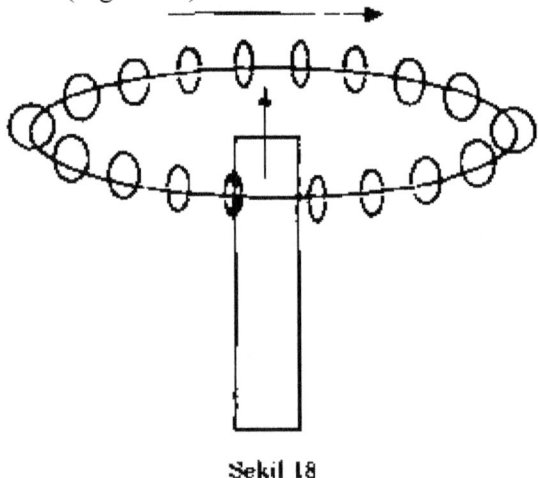

Şekil 18

Figure 18

If I approach the ring with a magnet from far away and very quickly with its axis in the middle, there is a change in flux and the charges start to rotate, just like in the wire. If this disc were on a wheel, it would have started spinning as soon as the magnet reached it. This seems to contradict the conservation of angular momentum; for when the magnet was away from the disc nothing rotated, but as it got closer it began to rotate. We got rotation without giving anything away; this is against the rules. They'll say to me, "Yes, I know, there must be some other interaction turning the magnet in the opposite direction." That's not the case. There is no electric force in the magnet to turn it in the opposite direction. The explanation for this is that angular momentum occurs in two ways. One is angular momentum of motion. The second is in the form of angular momentum of electric and magnetic fields. There is also angular momentum in the vicinity of the magnet; However, this is opposite to rotation and does not appear as motion. Looking at the opposite, the situation becomes clearer (Figure 19).

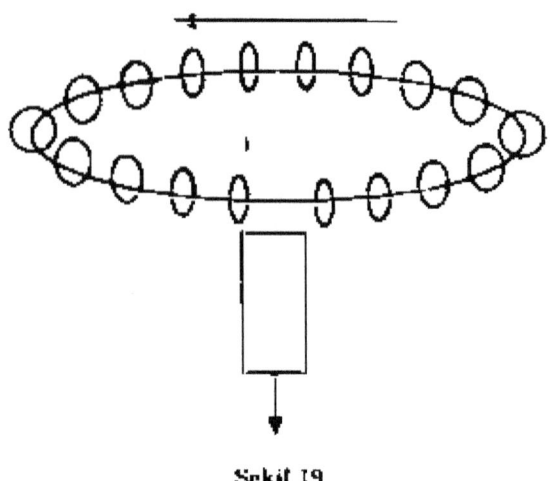

Şekil 19

Figure 19

When the magnet and the particles are close together and everything is at rest, there is angular momentum in the field; I would say there is latent angular momentum that does not act in the form of rotation. If you pull the magnet down and separate the system, all fields are separated and angular momentum must show; the disk starts to rotate. The law that turns the disk is the law of induction of electricity.

The question of whether the torque occurs in units is difficult for me to answer. At first glance it seems absolutely impossible that angular momentum occurs in units. Because the angular momentum depends on which direction we take the projection of the image. You are watching a field exchange; It's obvious that this changes depending on whether it's viewed from the opposite direction or from an angle. If angular momentum is in units, and if you look at something it says 8 units, then if you look at it from a slightly different angle, the number of units makes little difference, a little less than 8. However, 7 makes no difference "minor" out of 8; A certain amount is less than 8. So it cannot come in units.

we see that it always consists of a certain number of units. This unit is not a countable unit like electric charge. Torque appears as a specific integer times a unit for each measurement, which is units in the mathematical sense. However, we cannot interpret it as countable units like "one, then next, then again" as with electric charge. Angular momentum doesn't appear as single units, but strangely always as a whole number.

There are also some other conservation laws; but they're not as interesting as what I've described, and it's not exactly about conservation of numbers. Let's imagine an order with particles moving in a certain symmetry and assume that their motion is bidirectionally symmetric (Figure 20). After all the movements and collisions that happen according to the laws of physics, after a while you expect that everything is still symmetrical in both directions.

Figure 20

Here, too, a kind of preservation; The preservation of the symmetry property is

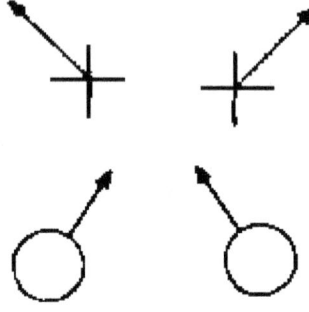

Şekil 20

questionable. This should also be shown in the table; However, it is not a number that we can measure. I will go into more detail on this in my next presentation. This is not an interesting, important, or practical feature for classical physics, since cases starting with such beautiful symmetries are rare. However, in quantum mechanics, which deals with very simple systems such as atoms, mostly the structures of systems

There is a symmetry similar to directional symmetry and the symmetry property is retained. Hence, it is an important law for understanding the quantum phenomenon.

Another interesting question is whether there is a deeper basis for all these conservation laws or whether we should accept them as they are. I will address this question in my next talk, but there is one important point I want to address now. If we discuss all of these ideas at a level understandable to all, it seems that there are many unrelated concepts. But when these various principles are understood more deeply, it becomes clear that there are deep interrelationships between them and that each somehow evokes the other. As an example, we can cite the relationship between relativity and local conservation. If I had said without explanation that the principle that we cannot know the speed at which we are moving,

Now I'm going to go into how conservation of angular momentum, conservation of momentum and some other things are related to some extent. Angular momentum refers to the area swept by moving particles. If there are many particles (Figure 21) and you take the very distant (x) as the center point, the distance for all the particles will be almost the same.

Figure 21

In this case, the only thing affecting the scanned range or the conservation of angular momentum is the vertical, which can be seen in Figure 21.

X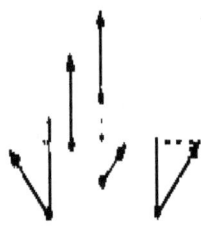

Şekil 21

part of the movement. Then the sum of all masses multiplied by their vertical velocity must also be constant. Since the angular momentum is constant around a point and the chosen point is very far away, only the masses and velocities change. Conserving angular momentum means that momentum is also conserved. This implies something else, the preservation of something else. Since these protected things are very closely related, I saw no need to include them in the table. This is a principle related to the center of gravity (Figure 22).

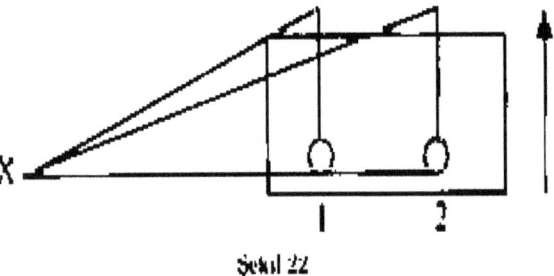

Şekil 22

Figure 22

A mass in a box does not disappear at one point and move to another point by itself. This has nothing to do with conservation of mass;

mass still exists; but he went from one place to another. Electric charge can do this, mass cannot. Let me explain why. The laws of physics are not affected by movement. Now suppose the box is slowly moving up and calculate the moment about a not very distant point x. If the mass stays at point 1 inside the box while the box is being pulled up, it creates a field with a certain speed. As the mass goes to point 2, the area increases rapidly. Because when the box is pulled up, the distance from x to the mass has increased, although the height remains the same. Since angular momentum is conserved, you cannot change the growth rate of the field. Because, if you don't balance the torque with some other thrust, you can't move the mass from one place to another. This is why rockets can't fly in a vacuum...but they do!

If you think about it by looking at a lot of masses, you will see that if you bring one forward you have to take the others back. Thus, the sum of the forward and backward movements of the masses becomes zero. This is how the rocket works. At first it stands motionless in space; Then he sprays gas from behind and goes forward. It is important that the center of gravity, the average of the total mass, stays where it was before. The interesting part has gone forward, the uninteresting part that doesn't interest us has gone backward. There is no theorem that says interesting things are preserved in the world; only the sum of everything remains.

Discovering the laws of physics is like putting the pieces of a puzzle together. There are many parts and their number is increasing rapidly these days. Many of them don't fit in with the others and hang around. How do we know they are parts of the same thing? We are indeed concerned about the fragments of a single unfinished painting; However, the similarities of some parts give us hope. Those showing the blue sky, or those made of the same species of wood... All the different laws of physics obey the same conservation principles.

In the laws of physicssymmetry

Symmetry is fascinating to the human brain. We enjoy looking at symmetrical things in nature, perfectly symmetrical spheres like the sun and planets, symmetrical crystals like snowflakes, almost symmetrical flowers. However, what I am going to discuss here is not the symmetry of objects in nature, but the symmetry of the laws of nature themselves. It is easy to understand whether an object is symmetric; but how can a physical law be symmetrical? Of course it can't. However, physicists love to use everyday words differently. With symmetry, they felt something similar to the feeling of symmetry in objects for the laws of physics and called it symmetry of laws. What is symmetry? If you look at me, I look symmetrical (right and left), at least from the outside. A vase can also be symmetrical in the same or a different way. How can you define it? My right-left symmetry means that if you transfer something from one side to the other and swap the two sides, the view stays exactly the same. The square has a special symmetry; If I rotate it 90° it looks the same. Mathematician Professor Weyl[17]He gave a very nice definition of symmetry: if, after manipulating an object, the object appears in its original state, if there is something in the object that allows this, then that object is said to be symmetrical.

We say that the laws of physics are symmetric in this sense. We can change some things about the laws of physics or the way they are expressed that don't change them in any way. In this lesson we will consider this aspect of the laws of physics.

The simplest example of this type of symmetry is displacement in space. You'll see it's not the left-right symmetry you're imagining or anything. Let's explain it this way; If you make a device or experiment with some things, and then make the same device or experiment with the same things, but not there but here, just moved from one place to another, then what happened in the original experiment is exactly that what happened in the postponed experiment. That's actually not quite right. If I build a gadget and move it 10 meters to my left, the gadget hits the wall and it gets tricky. When explaining this concept, it is necessary to consider everything that can affect the situation, and to convey everything together when conveying something. For example, if the system has a pendulum and I move it 20,000 miles to the right, the system will not work properly; because the pendulum is related to the gravitational force of the earth. However, if I transport the world along with the equipment, then the behavior of the system is not affected. The problem with this is the obligation to transport anything that could affect the situation. It might sound a little silly; It's like after postponing an experiment and getting a negative result, you can assume we didn't have everything necessary - so you're always right. The truth is not so; because you don't necessarily have to win. The interesting thing about nature is that it is possible to transplant enough things to behave in the same way. This is a positive statement. You can assume we don't have everything you need - so you're always right. The truth is not so; because you don't necessarily have to win. The interesting thing about nature is that it's possible to transplant enough things to behave the same way. This is a positive statement. You can assume we don't have everything you need - so you're always right. The truth is not so; because you don't necessarily have to win. The interesting thing about nature is that it's possible to transplant enough things to behave the same way. This is a positive statement. that we don't have everything you need - so you're always right. The truth is not so; because you don't necessarily have to win. The interesting thing about nature is that it's possible to transplant enough things to behave the same way. This is a positive statement. You can assume we don't have everything you need - so you're always right. The truth is not so; because you don't necessarily have to win. The interesting thing about nature is that it's possible to transplant enough things to behave the same way. This is a positive statement. that we don't have everything you need - so you're always right. The truth is not so; because you don't necessarily have to win. The interesting thing about nature is that it's possible to transplant enough things to behave the same way. This is a positive statement. You can assume we don't have everything you need - so you're always right. The truth is not so; because you don't necessarily have to win. The interesting thing about nature is that it's possible to transplant enough things to behave the same way. This is a positive statement. This is a positive statement. You can assume we don't have everything you need - so you're always right. The truth is not so; because you don't necessarily have to win. The interesting thing about nature is that it's possible to transplant enough things to behave the same way. This is a positive statement. This is a positive statement. You can assume we don't have everything you need - so you're always right. The truth is not so; because you don't necessarily have to win. The interesting thing about nature is that it's possible to transplant enough things to behave the same way. This is a positive statement.

 I want to show the truth of what I'm saying. Take the law of gravitation as an example, which states that the force between objects varies inversely with the square of the distance between them. Let me remind you that an object responds to a force by changing its speed over time and in the direction of the force. Let's take two objects like the sun and a planet revolving around it; If I translate both, the distance between them doesn't change, neither do the forces. Also, they move at their old speeds in their new positions, all changes happen at the same speed, and everything in both systems moves the same way. The fact that the law refers to the "distance between two objects"

So our first symmetry is displacement in space. We can describe the second as delay in time, or better yet, "the delay in time makes no difference". Let's move a planet around the sun in a specific direction. If we reactivate it two hours later or two years later at another time under the same conditions, it will behave exactly the same; because the law of gravitation speaks of velocity, but it says nothing about the absolute point in time at which we start measuring. In reality, we're not sure if this example is entirely accurate. In studying gravity, we talked about how the force of gravity can change over time. This possibility means that the time delay cannot be symmetrical. For if billions of years from now the gravitational constant will be weaker than it is now, it cannot be true that the motions of our experimental sun and planet will be the same billions of years from now.

I would love to be able to examine it as closely as possible!) a time delay does not lead to any change.

We know that in a way this is not entirely true; but true for what we today call the laws of physics. One of the facts of Earth is this: The universe seems to have started at a certain point in time and everything exploded. You could call it a geographic condition; It's a similar situation, when I shift in space, I have to shift everything. In the same vein, one could say that the laws of time are the same and that we should postpone the expansion of the universe along with everything else. We could also do another analysis where we could start the universe later. But we don't start the universe, we have no control over the situation, and there is no way to describe this idea empirically. So we can't be scientifically sure. The gist of this is: Earth's conditions seem to change over time; galaxies are moving away from each other; and if you wake up at an unknown time in a sci-fi story, you can measure the average distance to galaxies to tell the time. This means that the earth will not look the same with a time lag.

Today it is customary to separate the laws of physics, which describe the motions of bodies set in motion under certain conditions, from statements about how the earth actually formed; because we know very little about it. The history of astronomy or cosmology is believed to be slightly different from the laws of physics. But if I had to tell you what the difference is, I'd be in a difficult position. The best feature of the physical law is that it is universal; and if something is "universal" then it is the emanation of all star clusters. Therefore there is no way to define the difference. I don't know the method. However, if we leave the beginning of the universe aside and consider only the known laws of physics, I can say that the time lag makes no difference.

Let's consider more examples of the symmetry laws. One of them is rotation in space, constant rotation. If, after experimenting with some hardware installed in one place, we only get an exactly similar one with different axes (we may have to translate it so that it doesn't get on our feet), it works the same way. Again, we have to return everything relevant. If it is a pendulum wall clock and we turn the clock horizontally, the pendulum will touch the wall of the cabinet and the clock will not run. But if you also rotate the earth (it rotates constantly anyway), the clock continues to run.

The mathematical expression of this "spin opportunity" is quite interesting. When describing what is going on in a specific situation, we use numbers to indicate where something is. These are called the coordinates of a point. Sometimes we use three numbers to indicate how far a point is from a plane, how far forward (using negative numbers on back), or how far to the left. As such, I don't need to think up and down; because for the rotation I only need two of these three coordinates. Let x be the distance in front of me and y the distance to the left. Then I can indicate the position of an object by saying how far forward and how far left it is. People in New York City say house numbers work just as well (or just as well,
) You know. The mathematical approach to rotation is as follows; (Figure 23) If I determine the position of a point by giving the x and y coordinates using the method I

mentioned;
another looking from the other direction will also determine the position of the same point as x' and y', but relative to his own position. That's the mathematical expression. You write the equations with a few letters; There is a method of replacing the letters x and y with x', which is another x, and y', which is another y, that is, there are formulas related to x and y. Then the equations look the same, only the letters have a "'" on them. That is, the other sees the thing the way I see it, just the other way around.

    I will give another very interesting example of the symmetry law: the problem of uniform velocity along a straight line. It is believed that the laws of physics do not change at constant speed along a straight line. This is called the principle of relativity. Let's say we have a spaceship and a piece of hardware that comes with some functionality. If the spacecraft is traveling at a steady speed and someone inside is watching what is going on with the device, what they see can be nothing but what I see on my device on the ground. However, it's different if it's facing outside or hitting an outside wall or something like that; but the laws of physics will appear to him exactly as they appear to me, as long as he is traveling in a straight line at constant speed. So I don't know who's moving.

Before proceeding with this topic, I need to emphasize a few things. We're not talking about the shifting of the entire universe in all these transformations and symmetries. As for time, I'm not talking about the case that all time in the entire universe is shifted. So I'm not saying anything to say that if I put it somewhere else, everything in the universe will behave the same way. Importantly, if I buy a device and change its location; then I make sure that all necessary equipment is purchased and all conditions are met; I can take a piece of earth and move it relative to the average of all other stars; it still makes no difference.
relativity
In this regard, we can explain it as follows: a person moving along a straight line at a constant speed relative to the average of all star clusters observes no effect. In other words, you can't tell by any experiment whether you're in a car moving relative to all

the stars without looking outside.

This phrase was first uttered by Newton. Let's take his law of attraction. The law states that forces are inversely proportional to the square of distance and that force causes a change in velocity. Suppose I knew what happens when a planet moves around a fixed sun. Now I want to know what happens when a planet orbits a sun and travels through space. All the speeds I found in the first case change in the second case; I need to add a constant speed. However, the law is expressed in terms of changes in velocity. Since the force acting on the planet from such a fixed sun is equal to the force acting on the planet from the moving sun, the change in velocity of both planets will be equal, the additional speed that the second planet had at the beginning remains, and all changes are added. The mathematical result of this is this: if you add a constant velocity, the laws don't change, so we can't tell if the sun travels through space by studying the solar system and the motions of the planets orbiting it. According to Newton's law, such a trajectory does not affect the movements of the planets around the Sun. Therefore Newton added: "Although space is stationary with respect to some fixed stars, or moving along a straight line with constant velocity, the motion of the bodies relative to one another in space is the same." Over time, other laws were discovered after Newton. Maxwell's According to Newton's law, such a trajectory does not affect the movements of the planets around the Sun. Therefore Newton added: "Although space is stationary with respect to some fixed stars, or moving along a straight line with constant velocity, the motion of the bodies relative to one another in space is the same." Over time, other laws were discovered after Newton. Maxwell's According to Newton's law, such a trajectory does not affect the movements of the planets around the Sun. Therefore Newton added: "Although space is stationary with respect to some fixed stars, or moving along a straight line with constant velocity, the motion of the bodies relative to one another in space is the same." Over time, other laws were discovered after Newton. Maxwell's[18]

Among them are the laws of electricity that he discovered. One of the consequences of the laws of electricity was: Tamitamina per second.

Waves traveling 186,000 miles, electromagnetic waves

-Light is an example- it should have existed. By that I mean the speed is 186,000 miles per second in any case. Now the rest was easy to say; for the law that says the speed of light is 186,000 miles per second suggests (on the face of it) that it cannot be a law allowing motion without effect. If I stood still and you were traveling in any direction in the spaceship at 100,000 miles per second, if I sent you a beam of light traveling at 186,000 miles per second through a tiny hole in your ship; you per second

Since you travel 100,000 miles and light travels 186,000 miles, the speed of light appears to be 86,000 miles per second as it passes through your ship. However, if you do this experiment, the light will give you a second.

To me it looks like it's going at 186,000 miles per second.

It will feel like driving at 186,000 miles!

It is not easy to understand the phenomena in nature. The truth revealed by this experiment is so contrary to everyday logic that there are still those who do not believe this result! However, repeated experiments have shown that no matter your speed, the speed of light is always 186,000 miles per second. How can that be? Einstein and Poincare[19] They found that it would be possible for a standing person and a moving person to find the same result when measuring speed, only when a clock on a spaceship was running at a different speed than on Earth, because of their concepts of space and time so were different. "But when the clock is running slow, if I look at the spaceship's clock, I see that it is running slow," you might say. No, your brain is slow even on the spaceship! Everything is really like that in the spaceship

A system can be engineered so that a spacecraft appears at 186,000 my-miles per second inside the ship and at 186,000 my-miles per second on the ground. This requires great skill; And as hard as it is to believe, it turns out it's possible.

I have already mentioned a consequence of the principle of relativity; You don't know how fast you're going straight ahead.

If you recall, in our last lesson we had two vehicles, A and B (Figure 24), and there was an event on either end of that vehicle. A man was standing in the middle of the vehicle and the events (x and y) had occurred simultaneously at both ends of the vehicle at a given point in time. The man said that the events happened at both ends at the same time; Being in the middle of the vehicle, he saw the light at both ends at once. On the other hand, the person in vehicle A did not see the two events simultaneously since the light in x preceded the light in y as he was moving forward with respect to B at a constant speed; saw x first. As you can see, the word symmetry means symmetry when it comes to

One of the consequences of the principle - it happens that one cannot know who is seeing what is true - is this: "now" has no meaning when we are talking about what is happening "at the moment" in the world. As you move along a straight line at constant speed, events that seem to be happening simultaneously to you are different from events that seem to be happening to me at the same time; even if we pass each other at the time of the simultaneous event. When there is a distance between us, we cannot agree on "in the moment". This means that accepting the principle that constant-velocity motion along a straight line cannot be known requires profound changes in our conceptions of space and time.

I think you can see that this is very similar to x and y work in space. When I face the audience, the two walls of the stage I'm standing on are level with me; Their x's are the same, but their y's are different. If I turn 90° and look at the same two walls from a different angle, one is in front of me and the other is behind me and their x's are different. Thus, two phenomena that appear simultaneously from one point of view (same t) can appear at different times (different t') from another point of view. The two-dimensional rotation discussed previously has therefore been generalized to space and time, adding time to space to create a four-dimensional world. This is how many popular books say "we add time to space; because we cannot determine a point by just where it is, we must also say when it is" is not an artificial addition. What is said is true;

To a certain extent, real space has the property of having an existence that is independent of perspective. Viewed from different perspectives, "forward-backward" can sometimes be confused with "right-left". Likewise, some time, "past-future", can be mixed with some space. Space and time must be fully intertwined. After this discovery, Minkowski said: "Space itself and time itself are shrouded in darkness; only a kind of union between the two persists," he said.

 The reason I elaborate on this particular example is because they pioneered the study of symmetry in the laws of physics. It was Poincaré's idea to investigate what can be applied to equations without changing them. It was also his idea to give meaning to the symmetries of the physical laws. translation in space, time delay, etc. such symmetries are not very deep; but the symmetry of constant velocity on a straight line is very interesting and has many consequences. In addition, there are opportunities to extend these results to new laws. For example, if we estimate that this principle also applies to mumeson decay, we can say that with mu mesons we cannot know at what speed we travel in a spaceship. So while we don't know why mumezons spread, we can at least learn something about them.

There are also some symmetries of many different kinds. I will name just a few. One of them is this: It is possible to replace an atom with another atom of the same kind and it makes no difference as far as the facts are concerned. Now tell me, "What do you mean by the same kind?" You can ask. And I could answer that putting one in the place of the other makes no difference! physicist, a balama,

It seems like they always say nonsense, right? There are many different types of atoms; it makes a difference if you put different kinds of atoms in the place; but it makes no difference if you put the same type. It's a bit of a round definition going back to where it started. Its literal meaning is this: Atoms of the same kind exist, and it is possible to find a group, a class of atoms whose interchange produces no change. Considering that there are as many atoms as can be obtained by writing 23 zeros to the number 1 in the smallest piece of matter, it is important that they are all the same. It is indeed very interesting that they can be classified into a limited number of atomic types, for example a few hundred. Therefore, the meaning of saying that an atom can be replaced by an atom of the same kind is quite broad. The most far-reaching possibility lies in quantum physics. However, it is impossible for me to explain this problem now; Partly, but only partly, this is because this course caters to an audience that has not studied mathematics. The subject is very deep. The ability to replace one atom with another of the same type has wonderful results in quantum physics. Liquid helium also causes a strange phenomenon: liquid flowing through a pipe flows forever without resistance. It's actually the origin of the periodic table of elements, and that's why I didn't go through the base. I cannot go into the details here. I just want to emphasize the importance of exploring these principles. The subject is very deep. The ability to replace one atom with another of the same type has wonderful results in quantum physics. Liquid helium also causes a strange phenomenon: liquid flowing through a pipe flows forever without resistance. It's actually the origin of the periodic table of elements, and that's why I didn't go through the base. I cannot go into the details here. I just want to emphasize the importance of exploring these principles. The subject is very deep. The ability to replace one atom with another of the same type has wonderful results in quantum physics. Liquid helium also causes a strange phenomenon: liquid flowing through a pipe flows forever without resistance. It's actually the origin of the periodic table of elements, and that's why I didn't go through the base. I cannot go into the details here. I just want to emphasize the importance of exploring these principles.

At this point I think you've come to the conclusion that the laws of physics are symmetrical at every change. I will now give some examples where this is not the case. The first is a scale change. They made a device; then all parts are equal
It's not true that if you make a similar one, from the same material but twice the size, it will work exactly the same, exactly the same. You who are familiar with atoms know this. Because if you shrink the device, say, ten billion times, only five atoms remain inside; and I can't make a mechanical tool with five atoms, for example. It is obvious that going so far we cannot change the scale. But even before the concept of the atom was fully developed, it was clear that this law was not true. Every now and then one reads in the newspapers that someone has built a cathedral out of matches - a multi-storey cathedral, more "Gothic" and more elegant than any Gothic cathedral. Why don't we try Building cathedrals like this out of thick logs, tall and just as

elaborate and elaborate? The answer is: if we do something like that, it gets so high and heavy that it collapses. But we forgot something again. If you compare two things, we have to change everything in the system. The little cathedral of matches is pulled to the ground. Even the great cathedral must be drawn to a larger world. What a shame! A bigger world attracts more and trash breaks more easily. Even the great cathedral must be drawn to a larger world. What a shame! A bigger world attracts more and trash breaks more easily. Even the great cathedral must be drawn to a larger world. What a shame! A bigger world attracts more and trash breaks more easily.

Galileo was the first to discover that the laws of physics do not change when scale is changed. In discussing the durability of bones and rods, Galileo argued that for a larger animal -- say, twice as wide, tall, and thick -- the larger bone would be required, the weight eight times, and one bone eight times stronger. The load a bone can bear depends on its cross-section; If you increase the size of the bone twice, its cross-sectional area increases fourfold, and it pulls only four times as much weight.

In Galileo's Dialogue on Two New Sciences you can see images of extraordinarily large imaginary bones of giant dogs. I think Galileo realized that discovering that the laws of nature are not invariant with changes in scale was just as important as the laws of motion. Because he includes both together in his book Two New Sciences.

Another example that doesn't have a law of symmetry is this: it's not true to say that if you're spinning in a spaceship with constant angular velocity, you can't know whether you're spinning or not. because you know I can tell you're getting dizzy. There are other effects too; Under the influence of centrifugal force (perhaps there is another way of putting it - I hope there isn't an introductory physics teacher in the audience to correct me!) objects are thrown around. The rotation of the earth can be detected with a pendulum or gyroscope. Foucault, who proved in many observatories and laboratories that the earth rotates without looking at the stars.[20]You probably know it's a pendulum. We can say that on Earth we rotate at a constant angular velocity without looking outward. For under such motion the laws of physics are not immutable.

Many people have argued that the earth actually rotates relative to the galaxies and if we rotate the galaxies there will be no difference. I don't know what will happen if you rotate the whole universe; We currently have no way of determining this. We also currently have no theory to explain the effect of galaxies on the objects here. If so, it would be possible to understand that the inertia of rotation, the effect of rotation, the concavity of the water surface in a rotating bucket would have been caused by this theory - not by force or deceit, but directly - into our environment.

We could conclude that they are the result of the force exerted by the objects. We don't know if that's true. That this should be so is known as Mach's principle; however, this has not yet been proven. A more obvious empirical question is whether we see an effect when we rotate at a constant rate relative to star clusters. The answer is positive. If we travel in a straight line in a spacecraft at constant speed relative to the star clusters, do we see an effect? The answer is negative. These two are different things. We cannot say that all movements are relative. That is not what the theory of relativity contains. The theory of relativity tells us that it is unknowable to move in a straight line at constant speed relative to star clusters. Now I am going to tell you about a law of symmetry that is interesting in itself and in its history: the problem of reflection in space. I built a device, let's say a clock. Then I made a second watch, a mirror image of the first, a little further away. Equivalent to two gloves, left and right; a spring that turns one way in one direction, the opposite direction in the other, and so on. I set up two clocks, face each other and let them tick. The question is whether the two will always be compatible with each other. Does the entire mechanism also work in reverse in an existing mechanism? I don't know what your guess will be on this matter. Perhaps you will think that the answer is in the affirmative; The majority thought so. Of course we are not talking about geography. In geography we can distinguish right and left. If we stop in Florida and look at New York, we can say that the sea is on our right. This determines left and right. If the watch works with sea water, if we put it somewhere else, the watch will not work because it will not get water. We would then have to assume that the geography of the earth was also turned "on the other side"; everything related that the sea is on our right. This determines left and right. If the watch works with sea water, if we put it somewhere else, the watch will not work because it will not get water. We would then have to assume that the geography of the earth was also turned "on the other side"; everything related that the sea is on our right. This determines left and right. If the watch works with sea water, if we put it somewhere else, the watch will not work because it will not get water. We would then have to assume that the geography of the earth was also turned "on the other side"; everything related should be turned "to the other side". We are not connected to the past. If you pick up a screw in a turning shop, it is very likely that its thread is right-hand. You can argue that the other watch cannot be the same, that such a bolt is hard to find. It's just a matter of the things you do. The first guess is that it probably didn't notice anything. The result is this: if the clock were powered by gravity, the laws of gravity would be such that nothing would change and the clock would run. The laws of electricity and magnetism are the same; Clock, including electricity and some connections, currents, wiring etc. If so, the second clock would still work. If a normal nuclear reaction was required for the clock to work, that wouldn't make a difference either. But there is something that makes a difference; I will come to him shortly.

You may have heard that the concentration of sugar in water can be determined by passing polarized light through water. If you put a piece of polaroid in the water that only lets light through on a certain axis, you have to keep turning the polaroid material at the other end to the right to allow the light to pass through the deepening sugar water. If you rotate the light going through the water the other way, the rotation will still be to the right. So there is a difference between left and right here. We can use sugar water and light in hours. Let's say we have a water tank and we let light through and we rotate the second piece of polaroid just enough to let the light through. Then, for our second hour, we set up the appropriate order of the first, hoping the light turns to the left. But the light won't turn left, it will turn right again and won't go through the water.

This is a very interesting result, at first glance it seems to be proof that the laws of physics are not symmetric. However, the sugar we use may have come from beets. Since sugar has a very simple structure, it can also be obtained from carbon dioxide and water in the laboratory, sometimes using multi-stage processes. If you try this chemically identical artificial sugar in every way, you'll find that it doesn't turn the light. Bacteria eat sugar. If you put bacteria in artificial sugar water, they only eat half the sugar. If you shine polarized light through the remaining water after the bacteria have finished eating, you will see the light rotate to the left. The explanation for this is as follows: A sugar is a complex molecule made up of a complex arrangement of atoms. If you build the same arrangement alternately right and left, the distances between all pairs of atoms are the same in both; the energies of the molecules are also the same; the same applies to all non-life chemical events. However, living beings see a difference between them. Bacteria eat one species and not the other. Sugars made from beets are all of the same type, all right-handed molecules; That's why they only direct the light in one direction. Bacteria can only eat such molecules. If we make sugars from simple gases that are not themselves asymmetric substances, we produce equal amounts of both types. When we put bacteria in this environment, they eat whatever species they can eat; the other stays. Therefore the light rotates in the opposite direction. Pasteur's reason for this is that they only direct the light in one direction. Bacteria can only eat such molecules. If we make sugars from simple gases that are not themselves asymmetric substances, we produce equal amounts of both types. When we put bacteria in this environment, they eat whatever species they can eat; the other stays. Therefore the light rotates in the opposite direction. Pasteur's reason for this is that they only direct the light in one direction. Bacteria can only eat such molecules. If we make sugars from simple gases that are not themselves asymmetric substances, we produce equal amounts of both types. When we put bacteria in this environment, they eat whatever species they can eat; the other stays. Therefore the light rotates in the opposite direction. pasteurs[21] As he discovered, it is possible to distinguish these two types if we look at crystals with a magnifying glass. We can definitely show that all of this makes sense. If we want, we can separate the sugar ourselves, without waiting for the bacteria. But the interesting thing is that bacteria can do this. This means that living processes are not subject to the same laws.

he comes? As it turns out, yes. It seems that there are many complex molecules in living things, and they all seem to have some kind of screw thread. The smallest molecules in living things are proteins. These have a corkscrew function and turn to the right. So much so that we can say that if we can do the same things chemically, and if we do them to the left instead of to the right, they won't work biologically; because they cannot adapt when encountering other proteins. A left-hand thread mates with a left-hand thread; but left and right don't match. Bacteria with a right-handed groove in their chemical structure can distinguish between "left-handed" and "right-handed" sugars.

How do you achieve that? Physics and chemistry can make both types of molecules; however, it cannot distinguish them. But biology can tell. Such an explanation seems plausible: A long, long time ago, when life was just beginning, a molecule accidentally arose and spread through reproduction, etc. For many years these strange-looking, bifurcated droplets chattered to each other... And we are no more than the descendants of these few original molecules. It was coincidence that these first molecules tended to take on such a shape. They had to be this or that, right or left. Then they multiplied and are still multiplying. This is comparable to screws in a workshop. They make right-hand twists with right-hand twists, etc. This fact,

To better test the question of whether the physical laws are always the same on the right and left, we can put it this way: with someone who lives on Mars or on another planet.

Suppose you have a telephone connection and you want to explain objects on earth to it. First, how will he understand our words? This is a question from Professor Morrison at Cornell.[22] has been extensively studied by The method he proposes is: "Tick, one; tick-tick, two; tick tick tick, three; etc." to start with. Marsh will pick up numbers in no time. Once you understand the number system, write out the atomic weights, the whole series of numbers representing the proportional weights, and then "hydrogen, 1.008", then deterium, helium, etc. you go on.. After looking at these numbers for a while, he will realize that the mathematical ratios are the same as the ratios of the weights of the elements, and he will understand that these names are the names of the elements With this method, you can gradually build a common language. Now the following problem arises; After he approached him, he said to you: "You guys are so good. Let's say I want to know what you look like too." When you say, "We're about six feet tall," "Six feet? How big is a foot?" he asks. Very simply, "Six feet is as long as seventeen billion hydrogen atoms." This is no joke. Considering we can't send samples and none of us can look at the same object, this could be a way of describing six feet to someone who doesn't have a scale Since the laws of physics aren't true to scale, we can use that knowledge to scale and further define ourselves: We're 6'1" and outwardly bipartisan, we see it that way off, we have forked extensions, etc. He tells us, "These are very interesting; but how does your inside look like?' We also give him hearts etc. and we say: 'Now put the heart on the left side.' But how do you tell him which side is left? "Six feet is as long as seventeen billion hydrogen atoms." This is no joke. Considering we can't send samples and neither of us can look at the same object, this could be a way to describe six feet to someone who doesn't have scales. If we want to tell him how tall we are, we can. Since the laws of physics aren't true to scale, we can use that knowledge to scale and further define ourselves: We're 6'1" and outwardly bipartisan; we look like this, we have bifurcated extensions, etc. He tells us, "These are very interesting; but how does your inside look like?' We also give him hearts etc. and we say: 'Now put the heart on the left side.' But how do you tell him which side is left? "Six feet is as long as seventeen billion hydrogen atoms." This is no joke. Considering we can't send samples and neither of us can look at the same object, this could be a way to describe six feet to someone who doesn't have scales. If we want to tell him how tall we are, we can. Since the laws of physics aren't true to scale, we can use that knowledge to scale and further define ourselves: We're 6'1" and outwardly bipartisan; we look like this, we have bifurcated extensions, etc. He tells us,

"These are very interesting; but how does your inside look like?' We also give him hearts etc. and we say: 'Now put the heart on the left side.' But how do you tell him which side is left? Considering we can't send samples and neither of us can look at the same object, this could be a way to describe six feet to someone who doesn't have scales. If we want to tell him how tall we are, we can. Since the laws of physics aren't true to scale, we can use that knowledge to scale and further define ourselves: We're 6'1" and outwardly bipartisan; we look like this, we have bifurcated extensions, etc. He tells us, "These are very interesting; but how does your heart look like?" We also give him a heart etc. and we say: "Now put the heart on the left side. ' But how do you tell him which side is on the left? Considering we can't send samples and neither of us can look at the same object, this could be a way to describe six feet to someone who doesn't have scales. If we want to tell him how tall we are, we can. Since the laws of physics aren't true to scale, we can use that knowledge to scale and further define ourselves: We're 6'1" and outwardly bipartisan; we look like this, we have bifurcated extensions, etc. He tells us, "These are very interesting; but how does your inside look like?' We also give him hearts etc. and we say: 'Now put the heart on the left side.' But how do you tell him which side is left? Since the laws of physics are not independent of scale, can we use this knowledge to scale and further define ourselves: we are 1.80 m tall and outwardly non-partisan; we look like this, we have bifurcated extensions, etc. He tells us, "These are very interesting; but how does your inside look like?' We also give him hearts etc. and we say: 'Now put the heart on the left side.' But how do you tell him which side is left? Since the laws of physics are not scale independent, we can use this knowledge to scale and further define ourselves: We are 1.80 m tall and outwardly bipartisan; we look like this, we have bifurcated extensions, etc. He tells us, "These are very interesting; but how does your inside look like?' We also give him hearts etc. and we say: 'Now put the heart on the left side.' But how do you tell him which side is left? 80 m tall and externally non-partisan; we look like this, we have bifurcated extensions, etc. He tells us, "These are very interesting; but how does your inside look like?' We also give him hearts etc. and we say: 'Now put the heart on the left side.' But how do you tell him which side is left? Since the laws of physics are not scale independent, we can use this knowledge to scale and further define ourselves: We are 1.80 m tall and outwardly bipartisan; we look like this, we have bifurcated extensions, etc. He tells us, "These are very interesting; but how does your inside look like?' We also give him hearts etc. and we say: 'Now put the heart on the left side.' But how do you tell him which side is left? 80 m tall and externally non-partisan; we look like this, we have bifurcated extensions, etc. He tells us, "These are very interesting; but how does your inside look like?' We also give him hearts etc. and we say: 'Now put the heart on the left side.' But how do you tell him which side is left? Since the laws of physics are not scale independent, we can use this knowledge to scale and further define ourselves: We are 1.80 m tall and outwardly bipartisan; we look like this, we have bifurcated extensions, etc. He tells us, "These are very interesting; but how does your inside look like?' We also give him hearts etc. and we say: 'Now put the heart on the left side.' But how do you tell him which side is left? "They are very interesting; but how does your inside look like?' We also give him hearts etc. and we say: 'Now put the heart on the left side.' But how do you tell him which side is left? Since the laws of physics are not scale independent, we can use this knowledge to scale and further define ourselves: We are 1.80 m tall and outwardly bipartisan; we look like this, we have bifurcated extensions, etc. He tells us, "These are very interesting; but how does your inside look like?' We also give him hearts etc. and we say: 'Now put the heart on the left side.' But how do you tell him which side is left? "They are very interesting; but how does your inside look like?' We also give him hearts etc. and we say: 'Now put the heart on the left side.' But how do you tell him which side is left? Since the laws of physics are not scale independent, we can use this knowledge to scale and further define ourselves: We are 1.80 m tall and outwardly bipartisan; we look like this, we have bifurcated extensions, etc. He tells us, "These are very interesting; but how does your inside look like?' We also give him hearts etc. and we say: 'Now put the heart on the left side.' But how do you tell him which side is left? which side is left? Since the laws of physics are not scale independent, we can use this knowledge to scale and further define ourselves: We are 1.80 m tall and outwardly bipartisan; we look like this, we have bifurcated extensions, etc. He tells us, "These are very interesting; but how does your inside look like?' We also give him hearts etc. and we say: 'Now put the heart on the left side.' But how do you tell him which side is left? which side is left? Since the laws of physics are not scale independent, we can use this knowledge to scale and further define ourselves: We are 1.80 m tall and outwardly bipartisan; we look like this, we have bifurcated extensions, etc. He tells us, "These are very interesting; but how does your inside look like?' We also give him hearts etc. and we say: 'Now put the heart on the left side.' But

how do you tell him which side is left?
shall we say? You will say: "Take the beet
sugar, throw it in the water, it will...". But
that also has a downside; no turnips out
there! Furthermore, even if the same
proteins arose during evolution on Mars, we
have no way of knowing if they formed
reverse helices; So we can't tell. After much
thought, we realize that we cannot do this
and conclude that it is impossible.

However, some experiments conducted
about five years ago revealed many new
mysteries. I won't go into detail. We faced
increasing difficulties and illogical
situations. Finally Lee and Yang[23]They
argued that the principle of left and right
symmetry - that is, the principle that nature
is the same for left and right - may not be
true and that this might help explain some
unknowns. To demonstrate this, they
proposed some more direct experiments.
Among the experiments that were carried
out I will mention only the shortest.

Imagine radioactive fragmentation
releasing an electron and a neutrino. This
example is the fission mentioned earlier,
where a neutron gives a proton, an electron,

and an antineutrino. There are many radioactive events where the charge in the nucleus increases by one and the electron is released. The interesting thing here is that when electrons come out, they spin around on themselves. If you measure this rotation you will see that it points to the left (seen from behind). This has a certain meaning: the electron resulting from the scattering always rotates in the same direction; the groove is left-handed. It's as if the weapon that ejects electrons during beta decay is a gun. This weapon can be slotted in two ways. There is an "outward" direction; You have the option to turn left. Experiments show that electrons are ejected from the left rifled gun. With this result, he picked up the phone to the Martian and said, "Listen, take a radioactive substance, a neutron, and look at the electron that is produced by beta decay. If the electron goes up as it exits, determine its direction of rotation. If this electron entered from our back, the direction of rotation would be to the left. That defines the left. The heart is there, too," we say. In this way it is possible to distinguish between left and right, thereby breaking the law that the earth is right- and left-symmetrical.

Now I want to talk about the relationship between conservation laws and symmetry laws. In the last lesson on the principles of conservation; Energy, moment, torque, etc. We talked about conservation. It is extremely interesting that there is a very deep connection between the laws of conservation and the laws of symmetry. Quantum mechanics alone is the best explanation for this connection, at least according to current understanding. I'll show you an example of that now.

If we accept that the laws of physics can be explained by a minimum principle, then if a law allows you to move all the material to one side, i.e. if it can be moved in space, then conservation of momentum must also exist. There is a close connection between the principles of symmetry and the laws of conservation; However, this connection requires an acknowledgment that the principle of the minimum exists. In our second lesson, we discussed one way of expressing the laws of physics by saying that an object travels from one place to another by trying different paths in a given amount of time. We define a set, which we call an action, using a word that could lead to misunderstandings. Action for different species If we calculate it, we see that for the path chosen by the object, this quantity is always smaller than the others. Expressing the laws of nature in this way states that action is the smallest of all possible paths for the path chosen by the body. Another way of saying something is smallest is to say that a slight change in the object's path makes no difference. Suppose you walk on hills - but properly shaped hills, because mathematical things equal smooth, smooth things - and you get to the lowest place. Then I would say if you take a small step forward the height doesn't change. A step you take when you are at the highest and lowest point is the firstYour height with approximation won't affect it. However, on a downhill slope, one step down or one step up in the opposite direction.

This is key to why a step doesn't matter when you're at the bottom. Because when he notices, you take a step in the opposite direction and you're down. This is key to why a step doesn't matter when you're at the bottom. Because when he does, you take a step in the opposite direction and you're down. Since you are at the bottom and cannot go any lower, your first approximation would be that one step makes no difference. Therefore, a slight change in the object's trajectory is, to a first approximation, an action in the sense of an action.

We see that it makes no difference. We draw a path from A to B (Figure 25). Now let's take this as another route: first we jump to the nearest C and from there We go to another point that we can call D following a path similar to the first one. Since point D is on this second path, it will be shifted by the same amount. We now discover that the laws of nature are such that the total action on the road ACDB is equal to the action on the first road AB, which to a first approximation is actual motion by the principle of the minimum. Let me say one more thing; The action along the first path from A to B is the same action that would be taken from C to D if the world had not changed as we moved everything beyond.

Because the difference between the two is that we just postponed everything. So if the principle of translational symmetry holds in space, the action on the first direct path from A to B is the same as the action on the direct path between C and D. However, for real movement: the action along the indirect ACDB is almost the same as the action along the direct EU. Therefore, the same applies to the direct CD partition. This indirect action is the sum of three separate parts: A to C, C to D, and D to B. Since the sum is zero when equal is subtracted from equal, the sum of AC and DB must also be zero. Movement in one of these segments is in one direction and in the opposite direction in the other. If we think of the contribution from A to C as moving in one direction and from D to B as the opposite sign from B to D, since the one from D to B goes in the opposite direction, then the number from A to C is B. It should be big enough to take from ' to '. This is the effect of a tiny step from B to D on the plot. This quantity, which is the effect of a small step to the right on the action, is the same as at the beginning (A to C) and at the end (B to D).

So if the principle of minimum is true and the symmetry principle of translation in space is true, then there is a time-invariant quantity. This constant amount (the effect of a small step to the side) so the effect of a small step to the right on the action is the same as at the beginning (A to C) and at the end (B to D). So if the principle of minimum is true and the symmetry principle of translation in space is true, then there is a time-invariant quantity. This constant amount (the effect of a small step to the side) so the effect of a small step to the right on the action is the same as at the beginning (A to C) and at the end (B to D). So if the principle of minimum is true and the symmetry principle of translation in space is true, then there is a time-invariant quantity. This constant amount (the effect of a small step to the side) is the momentum itself, which we examined in the last lecture. If we accept that the laws are governed by the principle of minimum action, this shows us the relationship between the laws of symmetry and the laws of conservation. From this it follows that the laws obey the principle of minimum action as they come from quantum mechanics. For this reason, I have already told you that the explanation of the relationship between the laws of symmetry and the laws of conservation occurs in quantum mechanics.

An analogy to this discussion for the time delay brings up energy conservation. Ignorance of rotation in space also suggests conservation of angular momentum. Not noticing the reflection does not lead to anything simple in the classical sense. This thing was called parity, and a conservation law called conservation of parity was introduced. But these are nothing more than complex words. I have to mention parity preservation for a reason; You may have read in the newspapers that the law of conservation of parity has been proven wrong. It would be easier to understand if this were written in a way that refutes the principle that left and right are indistinguishable.

As an aside, I'd like to address some of the new problems that have arisen regarding symmetries. For example, for every particle there is an antiparticle; For the electron it is the positron, for the proton it is the antiproton. In principle we can do what we call antimatter. Every atom in antimatter consists of antiparticles of atoms in matter. The hydrogen atom is one electron and one proton. If we combine an electrically negatively charged counterproton with a positron, a kind of hydrogen atom is formed, counterhydrogen.

The counter hydrogens are not actually "made"; However, it is believed that in principle they can be done, just as any kind of antimatter can be made. So does antimatter behave like matter? As far as we know, yes. One of the symmetry laws states that everything we do with antimatter behaves the same as what we do with matter. However, when they come together, they create sparks and annihilate each other.

Matter and antimatter were considered to be subject to the same laws. But we now know that right and left symmetry can be wrong. In this case, an important question arises. Neutron fragmentation is for an antimatter - an antineutron; Consider - a counterproton decays into a counterelectron (positron) and a neutrino. The question is: will it behave the same way, that is, will the positron come out with a left-handed helix? Or will it behave differently? Until a few months ago he acted in the opposite direction; We believed that when matter moved to the left, antimatter (positron) would move to the right. In this case we had no way of telling the Martian what was right and what was left; because if he were made of antimatter, would the martian put the heart on the opposite side when doing the experiment as its electrons would be positrons and they would spin in the opposite direction. Suppose you call the Martian and tell him how to make a human. He does what you say and he succeeds. Then you explain to him all our social customs. He also tells us how to build a good enough spaceship. Finally you go to him. You walk towards him and stretch out your right hand to shake his hand. If he also stretches out his right hand, all is well. But when he stretches out his left hand, watch out... you'll destroy each other! If he also stretches out his right hand, all is well. But when he stretches out his left hand, watch out... you will destroy each other! If he also stretches out his right hand, all is well. But when he stretches out his left hand, watch out... you'll destroy each other!

I would like to give you other examples of symmetry; but the explanations are getting harder and harder. There are some very interesting things that we call approximate symmetry. The remarkable thing about being able to distinguish right and left, for example, is that we can only do this with a very weak effect, this beta fragmentation. This is right and left in nature, percent 99.99 probably means they are indistinguishable from each other; but it also means that something completely different, upside down, tiny, a tiny phenomenon exists. This is an unfathomable secret that no one has the slightest idea of yet.

Distinguish between past and future

It is known to everyone that the phenomena of the world cannot be reversed. That is, something happens and they don't evolve back in the opposite direction. If you drop a cup on the floor, the cup will break; If you're waiting for the pieces to snap back together and squirt on your hands, you'll be waiting in vain. As you watch the waves of the sea break, you wait a long time for the big moment when the spray gathers again, rises from the sea, recedes further from the shore and rises again - how nice that would be!

The presentation of this topic in class is usually done by playing a movie backwards that contains some events and waiting for the laughter. That laugh is real It means that such things cannot happen in the world. However, that's not the best way to explain something as profound and obvious as the difference between past and future. For although we do not experiment, the experiences we have accumulated show that the past and the future are completely different. We remember the past, not the future. Awareness of what could have been is very different from awareness of what could have been. From a psychological point of view, future and past are different because they involve the concepts of free will and memory: we feel that we can do something to affect the future; but none of us think we can do anything to influence the past; Or few of us do. Regret, guilt, hope, etc. Words most clearly distinguish past and future.

If the realm of nature is made of atoms, and we are made of atoms and subject to the laws of nature, the most obvious explanation for the apparent difference between past and future and the fact that events cannot be undone may be as follows: some laws, like some Laws governing the motion of atoms work in one direction only; Atomic laws don't work either way. Somewhere there has to be a principle that "uxleys" only become "wuxleys" and not the other way around. So the world is always changing from uxleys to wuxleys; and this one-way interaction between objects causes the earth to always go in one direction.

However, we have not yet discovered this principle, and therefore we have not discovered any difference between past and future in all the laws of physics that we have discovered so far. From this point of view, the film should be able to work equally well in both directions; the physicist looking at him shouldn't laugh either.

Let's take our usual example, the law of gravity. Let me have a sun and a planet. Suppose I move the planet in one direction around the sun and make a movie of it, then I run the movie backwards and watch it. What happens? The planet moves in an ellipse around the sun, of course in the opposite direction. The speed of the planet is such that at equal time intervals the area swept by the radius is equal. So really, it moves like it should. It's no different than going in the opposite direction. Accordingly, the Law of Attraction is a law where direction makes no difference. If you just play the movie of a gravitational event in the opposite direction, everything seems fine. More precisely, one can put it this way: If the velocities of all particles in a complex system suddenly reverse, then the system dissolves backwards from what created it. If you have a bunch of particles doing something and you suddenly reverse speed, they will undo what they were doing before.

This is the law of attraction, which states that force changes speed. If I reverse time, the forces don't change and consequently speed changes are unaffected. Each velocity will then experience successive changes in a manner that is the exact opposite of the phases that formed them previously. It is easy to prove that the law of attraction is reversible in time.

What can we say about the laws of electricity and magnetism? These are reversible over time. What about the laws of nuclear interaction? As far as we know, they are also time-reversible. That beta decay law we talked about earlier? Are they also time reversible? Experiments a few months ago showed something was wrong.

that there is something in the law that is not understood; suggesting that beta decay may not be reversible over time. It is necessary to wait for new experiments to understand them. However, we can at least say that this is true: beta decay (whether time-reversible or not) is a trivial phenomenon for most normal situations. My ability to speak here is independent of beta decay, although dependent on chemical interactions and electrical forces; Though fairly independent of nuclear forces, it is dependent on gravity for now. But I'm "one-sided"; When I speak, a sound is emitted into the air, but when I open my mouth, that sound does not return and enters my mouth. Again, this inversion cannot be attributed to beta decay. In other words, we believe that most normal events that occur in the world as a result of atomic motion are governed by completely reversible laws. Therefore, we need to do more research to find an explanation for the non-reversal phenomenon.

If we take a closer look at the planets orbiting the sun, we quickly realize that all is not going well. For example, the Earth's motion around its own axis is gradually slowing down. This is due to tidal friction, and obviously friction is irreversible. If I put a heavy object on the ground and push it, it slides and stops. If I stop and wait it doesn't come back to my hand by suddenly getting up and speeding up. Thus, the frictional force appears to be irreversible. However, as we have already explained, a frictional force is a very complex thing, arising from the interaction between the weight and the wood and the vibration of the atoms inside. The orderly movement of the body, the random irregularity of the atoms on the board. turned into movements. We need to examine that more closely.

Here we have a clue to illuminate the phenomenon of non-reversal. I'll take a simple example. Let's put blue colored water on one side of a partitioned water tank and uncolored water on the other side; then let's remove the partition very slowly. In the beginning the waters are separated. Let's wait a bit. Gradually, the blue water mixes with the colorless water, and after a while the color mixes properly, making the water "bluish". That is, it spreads evenly halfway through the blue tank. Then, even if we wait and watch for a long time, it doesn't go away on its own (you can do something to separate the blue. You're evaporating the water and condensing it somewhere else. You can mix the blue dye with half of it. But while you're doing all that, you're in a different place to cause a phenomenon that's irreversible. However, blue water does not go the other way on its own.

That gives us a clue. Let's look at the molecules. Let's say we're filming a mix of blue and white water. If we point the film in the opposite direction, we come across a rather strange image. Because once you started with monochromatic water, the colors gradually diverge. Now let's zoom in on the picture enough that physicists can see what's irreversible by looking at the atoms individually (where the laws of forward and backward equilibrium break down) and looking again and you'll see two types of atoms (which is ridiculous but we call them blue and white), and these atoms stay in thermal motion, they keep oscillating. To recap, often atoms of one kind are on one side and those of the other kind on the other. Billions upon billions of atoms, one kind on one side and the other on the other, vibrate, At the end of the movements we see that they are mixed together. Therefore, water has an almost uniform blue color.

Let's consider a single collision that we will select from this film. In the movie, atoms come together in that direction and bounce in that direction. Now if we reverse that part of the movie, we see the pair of molecules approaching in that direction and jumping in that direction. The physicist takes a close look and measures everything and says, "Okay, according to the laws of physics, when two molecules collide like this, they bounce off like this." This is twofold. The collision laws of molecules are reversible.

That means that if you look very closely, you can't understand anything; because each of the collisions is strictly reversible. But even what is seen in the film is strange. Because when the film is upside down, the molecules start out mixed—blue, white, blue, white, blue, white; During collisions, the blue is separated from the white. However, they cannot do this; It is not self-evident that blues differ from whites in real cases. However, if you look closely at the reverse movie, you will see that every collision is reversible.

From all of this you can see that reversibility is based on the accidents of life. If you start with something separate and make infrequent changes to it, that thing becomes more unified; but it doesn't decompose if you start with uniformity and make infrequent changes. It could separate. It does not violate the laws of physics for molecules to jump to dissociate; it's just very unlikely. It won't happen in a million years, so that's the answer. Even in a million years, events are only possible in one direction but in the other, although compatible with the laws of physics.

are irreversible in the sense that they cannot occur. It would just be ridiculous to sit in one place for a long time and wait for the vibrations of the atoms to separate the paint on one side and the water on the other in an even mixture of water and paint.

If I do this experiment with a box of four or five molecules of each type, over time the molecules will mix. I think if you look further, after a while - it doesn't have to be a million years, maybe just a year later - the molecules happen to be almost in their original state as a result of disordered disordered collisions, that is, if I have a partition between all with the blue on one side and all the white on the other side, you can wait for them to reach their original state. This is not impossible. However, the real objects that we use in experiments are not made up of four or five blue and white molecules. There are four or five million million million million million molecules and they all have to separate like that. The apparent irreversibility of nature is not due to the irreversibility of the fundamental laws of physics. Why is this:

So our next question is how they were ordered in the beginning. So how is it possible to start with "normal"? The difficulty with this is that you start with something regular and you end up with something regular. One of the rules of the world is that things go from regular to irregular. Incidentally, the words "regular" and "irregular" are physical terms that are not used in their ordinary meaning. You may not be interested in the layout. However, there is a certain situation here; everything is on one side and everything on the other, or they are mixed together; This is order and disorder.

So here's your end: How did they organize in the beginning? And if we look at a condition that is partially regular, why do we conclude that it may be due to something more regular? If I look at a water tank with dark blue side and clear white on the other side and bluish water in the middle and know that it has not been touched for twenty or thirty minutes, I would suspect that the reason this condition is due to it that he was completely separated in the past. The longer I wait, the more blue and white mix; While I know this thing hasn't been touched long enough, I can deduce something about its former state. Being "smooth" on the sides can be the result of that he was better apart in the past; because since then, if it hadn't been better separated earlier, it would have been a lot more confused than it is now. So if we look at the present, we can say something about the past.

In reality, physicists don't go that far. Physicists simply say that "the conditions are like this; What's going to happen now?" they tend to say it means. All the other sister sciences have a different problem. In fact, all other fields of study -- history, geology, astronomy, history -- have a different set of problems, and those who "I'll tell you what happens next under these conditions," said one physicist, while a geologist said, "Digging the ground, I found some bones. If you dig up the ground, you'll probably find bones of the same species," he says. The historian also talks about the past However, this is done by talking about the future. What he meant when he said the French Revolution took place in 1789 is that if you look at another book on the French Revolution you will see the same date. What he's doing is speculating about documents he's never seen, that haven't been found yet. He argues that the documents in which something is written about Napoleon will be the same as those in other documents. How is that possible? The only thing that makes this possible is the assumption that Earth's past was more orderly in that sense than its present state. Some people have suggested that the earth is fine, which I'll explain

below. In the beginning the whole universe consisted of irregular movements, as in colored water. I said that if there are very few atoms, water can be randomly separated if we wait a long time. Some physicists suggested (a century ago) that these were all just fluctuations (a term denoting a slight deviation from the normal regular state) of the world, a long-standing system. There have been fluctuations and we are now watching these fluctuations dissipate and return to the old state. "But imagine how long it takes to wait for a surge like that," you might say. I know, but we wouldn't notice if the fluctuation wasn't large enough to allow evolution to create an intelligent human being. So there had to be a flood of at least this magnitude where we had to wait until we could recognize them. Still, he thinks that theory is wrong, and that's why I find it absurd. If the world were much larger and had atoms in a complex state everywhere; if I just look at atoms in one place and see them separately
I would not have concluded that the atoms on the ground were also separated. If it was a wave and I saw something unusual, it was

probably there because there wasn't anything unusual anywhere else. In the white and blue water experiment, if some of the molecules in the box are separated, the most likely situation for the remaining water is that the colors are still mixed. Likewise, when we look at the stars and the earth, we see all is well; If there were fluctuations, we would expect to see chaos and disorder when we look where we haven't looked before. We have seen that matter is separated into hot stars and cold space. This may be due to staff turnover. Only then can we speculate that the stars have not left the space where we do not look. As we find the statement about Napoleon, or we can predict that we will see bones similar to those we have seen before, and we think that in places we don't look, there are always stars in a state will be similar to those we know. Achievements in all sciences indicate that the earth did not emerge from fluctuations, but from a more ordered and separate state in the past. Therefore I think it is necessary to add to the laws of physics the hypothesis that the universe was more technically ordered in the past than it is today. I think it is this statement that we need to add in order for irreversibility to make sense and be understood. shows, that he came from a situation that was more organized and disconnected in the past. So I think it is necessary to add to the laws of physics the hypothesis that the universe was more technically ordered in the past than it is today. I think it is this statement that we need to add in order for irreversibility to make sense and be understood. indicates he came from a situation that was more organized and disconnected in the past. So I think it is necessary to add to the laws of physics the hypothesis that the universe was more technically ordered in the past than it is today. I think it is this statement that we need to add in order for irreversibility to make sense and be understood. So I think it is necessary to add to the laws of physics the hypothesis that the universe was more technically ordered in the past than it is today. I think it is this statement that we need to add in order for irreversibility to make sense and be understood. indicates he came from a situation that was more organized and disconnected in the past. So I think it is necessary to add to the laws of physics the hypothesis that the universe was more technically ordered in the past than it is today. I think it is this statement that we need to add in order for irreversibility to make sense and be understood. So I think it is necessary to add to the laws of physics the hypothesis that the universe was more technically ordered in the past than it is today. I think it is this statement that we need to add in order for irreversibility to make sense and be understood. indicates he came from a situation that was more organized and disconnected in the past. Therefore I think it is necessary to add to the laws of physics the hypothesis that the universe was more technically ordered in the past than it is today. I think it is this statement that we need to add in order for irreversibility to make sense and be understood. which we need to add in order for irreversibility to make sense and be understood. indicates he came from a situation that was more organized and disconnected in the past. So I think it is necessary to add to the laws of physics the hypothesis that the universe was more technically ordered in the past than it is today. I think it is this statement that we need to add in order for irreversibility to make sense and be understood. which we need to add in order for irreversibility to make sense and be understood. indicates he came from a situation that was more organized and disconnected in the past. So I think it is necessary to add to the laws of physics the hypothesis that the universe was more technically ordered in the past than it is today. I think it is this statement that we need to add in order for irreversibility to make sense and be understood.

The movement itself is upside down in time; says something about the past is different than the future. However, this is usually outside the realm of the laws of physics. Because we no longer try to separate the expressions of the physical laws that govern the evolutionary rules of the universe from the laws that express how the earth was in the past.
we work. Now this is considered the history of astronomy; maybe one day in the future it will become part of the laws of physics.

Now I want to show you some

interesting properties of irreversibility. One of them can be seen by observing how an irreversible machine works. Let's say we're building a machine that only has to run in one direction. I'm making a circular saw shaped wheel with teeth curved vertically up and down all the way around. The impeller was set to a mile. There is also a rod (ratchet) attached to a knuckle by a spring, which can prevent the wheel from spinning backwards (Figure 26). The wheel can only turn in one direction.

Figure 26

If you try to turn it the other way, the pawl will catch the vertical teeth and won't

turn; when you turn it in the right direction, it rides over the teeth with a "click" sound (you've seen this in watches; there is this in wristwatches. You wind it in one direction;

after you wind it , the pawl holds the mainspring). The wheel is completely irreversible in the sense that it can only turn in one direction. opposite
such a machine that cannot be turned, that can be turned in one direction

Figure 27

We know that molecules are in constant motion. When you build a very sensitive instrument, the instrument will constantly

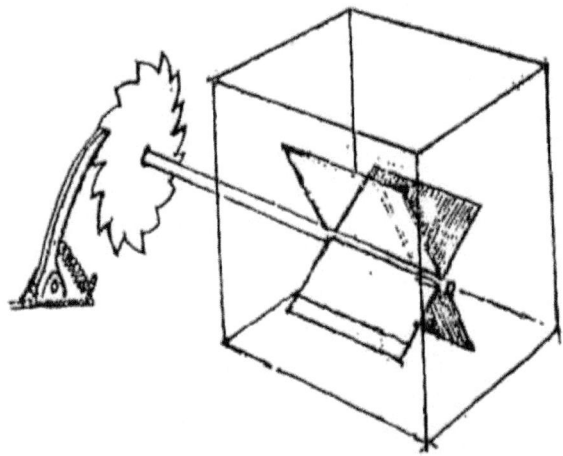

vibrate as it is subjected to constant, erratic bombardment from the surrounding air molecules. Now let's connect our wheel to a four-bladed propeller using the shaft on it, as shown in Figure 27.

The wings are in a gas-filled box and are constantly being bombarded by molecules; The wings are sometimes pushed in one direction and sometimes in the other. They are held by the pawl when pushed in one direction, and rotate when pushed in the other direction. So the wheel spins without stopping, and we get a kind of continuous motion. What makes this possible is the non-reversibility of the transmission.

We need to watch this happening more closely. The device works by the wheel lifting the pawl while rotating in one direction, and then the pawl falls, locking the tooth. The paddle goes up and down; When the spring is working properly, it goes up and down continuously, and if the pawl goes up somehow, the wheel spins in the opposite direction. If the pawl doesn't lock when it's down or doesn't go up and down, the mechanism isn't working. Friction or deceleration (gradual decrease in damping vibration) occurs during the up and down movement, which is the only thing that ensures rotation in one direction; The heating occurs due to friction and the wheel gets hotter and hotter and as it starts to heat up, another event occurs. Whatever material they are made of, when the wheel and pawl heat up, they cause the irregular motions of the gas around the blades, i.e. Brownian motions.[24] irregular movements in the wheel also begin. Over time, the wheel becomes so hot that the pawl begins to vibrate due to the molecular movements it contains; it bounces up and down due to molecular motion (the same effect that makes the wings spin). As the pawl bounces up and down on the wheel, it stays both down and up. Meanwhile, the tooth can go either way. Our device is no longer unidirectional; It can also work backwards! When the impeller is hot and the impeller part is cold, the impeller, which we think only moves in one direction, moves in the other direction. Because every time the pawl goes down, it falls onto the sloping surface of the teeth on the wheel, thus pushing the wheel "backwards". Then it rises again and falls onto the sloping surface of another tooth, and so on. That is, when the impeller is hotter than the blades, it moves in the opposite direction.

How does all this relate to the

temperature of the gas around the blades? Suppose this piece never existed.
If the wheel is pushed forward while the pawl falls on the rake of the tooth, the vertical face of the tooth will hit the pawl and the wheel will spring back. To prevent the wheel from jumping backwards, we install a retarder and send air to the wings; So the wheel slows down and doesn't jump free. Now it only goes in one direction; but in the opposite direction. Such a wheel spins in one direction when one part is hotter than the other, and in the opposite direction when the other is hotter. But when there is heat exchange between the two sides, that is, when the temperature is the same in both the blades and the impeller, it goes neither to nor there. As long as there is no equivalence,

    Saving energy can make us believe that we have as much energy as we want. Nature never loses or gains energy. But for example the energy in the sea, the thermal energy caused by the thermal movements of the atoms in the sea can be practically ignored for us. In order to regulate, direct and use this energy, a temperature difference must exist. Otherwise, although the energy is there, we cannot benefit from it. There is a big difference between energy and usable energy. There is great energy in the sea; however, it is not available to us.

Conservation of energy means that the total energy of the world does not change. The energy emitted by irregular vibrations can be so uniform that in some cases it is not possible to transfer more in one direction than the other to control it.

I think I can give you an idea of the difficulties in this matter through an analogy. Have you ever been caught by a sudden and heavy downpour on the beach with your towel next to you? it happened to me You quickly pack up the towels and run to the beach cabin, then you start drying them. The towel got a little wet; However, it is still drier than you. You dry with the towel until it gets wet, and you find that it wets you as much as it dries you. Then you get another one, but you soon realize that all the towels are as wet as you are. It doesn't get any drier when you have a lot of towels; because there is no difference between your wetness and the wetness of the towels. Now I can invent a size that says "extinguishing comfort" names. Towels are just as easy to water as you are. Therefore, if you touch yourself with a towel, the water flowing from you to the towel will flow from the towel to you. This does not mean that the towel holds the same amount of water as you do. There is more water in a large towel than in a small towel; however, their humidity is the same. Once the wetness of the objects is balanced, there is nothing you can do.

Water is like energy; because the total amount of water does not change (If the beach cabin door is open, you can run outside in the sun to dry, or find another towel to get rid of. However, we assume everything is closed and you can't get new ones buy towels). If we assume that part of the world is also closed and wait long enough; As a result of random events in the world, energy will be evenly distributed, as in the case of water, the one-way street will disappear, and there will be no trace of real interest in this regard. So narrow, which contains nothing else

As in the case of the encompassing wheel, pawl and propeller, the temperature on both sides gradually becomes the same and the wheel does not rotate in one direction or the other. Likewise, if a system is left alone for any length of time, the energy in it will become confused and there will be no energy left that can be used for any work.

Here it is temperature that corresponds to wetness or "ease of quenching". We say that equilibrium occurs when two things are at the same temperature, but that doesn't mean their energies are equal; it simply states that it is just as easy to extract energy from one as from the other. Temperature is something like 'ease of energizing'. If you put them side by side, nothing can be seen. They conduct energy back and forth evenly; however, the net result is zero. So if all objects reach the same temperature, there is no energy that we could use to do anything. The principle of irreversibility is that when objects are at different temperatures and left to their own devices,

This is a variation of the entropy law, which states that entropy is constantly increasing. Let's not dwell on words. In other words, we can say that the usable energy is constantly decreasing. This is a feature of the world caused by the chaos of disordered molecular movement. If things with different temperatures are left to their own devices, they tend to be the same temperature. If you have two things at the same temperature, for example water on an unlit stove, the stove will not heat up and the water will not freeze. However, the opposite happens when a stove and ice are on fire. Unidirectionality always leads to the loss of usable energy.

That's all I have to say about it. However, I would like to make a few points about some key features. Here we have an example quite different from the fundamental laws, the consequence of which is obvious, such as irreversibility, but not the obvious consequence of the laws. To understand why this requires a lot of analysis. This result is very important for the world economy and its actual behavior on every seemingly obvious issue. My memory, my features, the difference between past and future are all intertwined with it. However, knowing the laws is not enough to easily explain this; A lot of analysis is required.

It is a common situation that there is no obvious and direct correspondence between the laws of physics and the facts. Laws are abstracted from experience to varying degrees. In this particular case, for example, laws are reversible but facts cannot be reversed.

There is often a great distance between detailed laws and the essential characteristics of real facts. For example, viewing a glacier from a distance and viewing large rocks falling into the sea, ice movement, etc. You don't have to remember that it's made of tiny hexagonal ice crystals when you see it. But we do know that moving ice is actually caused by hexagonal ice crystals. It takes a long time to understand the behavior of the glacier (in fact nobody knows enough about ice, no matter how much they have studied the crystals). However, we hope that if we truly understand crystals, we will eventually understand glaciers as well.

Although in these lectures we are talking about the basic elements of the laws of physics, I should add that the basic laws of physics are If we know as much as possible today, we still don't understand anything. This takes time, although we can only partially understand it. It's as if nature had been arranged in such a way that the most important things in the real world seem like a jumble of random results of a set of laws.

An example: atomic nuclei are very complex and contain some nuclear particles such as protons and neutrons. They have what we call energy levels and they exist in states or states with different energy values. The energy levels of different nuclei also differ from each other. Determining the state of energy levels is a complex mathematical problem; We can only partially solve this. The exact state of the levels is the result of something extremely complex. Therefore, nitrogen with 15 particles in it would have a 2.4 million volt level, another 7.1 level, etc. It is not surprising that it is so. There is something very interesting about nature: the particular structure of the entire universe depends on the state of a certain energy level in a certain nucleus. It was determined,

Here's the situation: Let's start with

hydrogen. At first glance, the earth appears to consist almost entirely of hydrogen. Under the influence of gravity, the hydrogen is compressed and heated, and a nuclear reaction takes place; helium is formed. Then helium partially combines with hydrogen to form several heavier elements. However, these heavier elements immediately dissipate and turn back into helium. Therefore, it was not possible to understand how all other elements in the world came into existence. Because the production process in stars begins with hydrogen and extends beyond helium and less than half a dozen other elements.

couldn't tell. Faced with this problem, Hoyle and Saltpeter[25] They suggested that there was a way out. So if three helium atoms can combine to form one carbon, we can easily calculate how many times this can happen in a star. The result showed that carbon could only have been formed by a single random possibility. If carbon had an energy level of 7.82 million volts, the three helium atoms could combine and stay together a little longer than if the 7.82 level weren't there. Staying a little longer would allow time for something else to form and for new elements to be crafted. If in carbon If there were an energy level of 7.82 million volts, it would be understandable where the other elements on the periodic table come from. Thus, by indirect and inverse analysis, a level of 7.82 million volts was estimated for carbon; Laboratory experiments have shown this to be true. Therefore, the existence of all other elements in the world is closely related to the existence of that particular level of carbon. The existence of this special plane in carbon gives us who know the laws of physics the impression that it is a very complex random result of the interaction of 12 complex particles. This example is a very good example that understanding the laws of physics does not require a direct understanding of the important things in the world.

When we discuss the world, we look at it in a hierarchical order and at different levels. By that I don't mean dividing the world into definite and definite levels. I'll show what I mean by the hierarchy of ideas by explaining a number of concepts.

At one extreme, for example, are the basic laws of physics. We invent other concepts for approximations, which we think we can expect to be explained in detail by fundamental laws; eg "Temperature". We think temperature is vibration; The word we use for something hot is also the word we use for a vibrating mass of atoms. But when we talk about temperature, we forget vibrating atoms. Just like when we talk about glaciers, we forget about the hexagonal ice and snowflakes that fall in the first place.

Another example of the same is salt crystals. They essentially consist of many protons, neutrons and electrons. But we have the concept of a "salt crystal" that contains the entire basic order of interaction. Pressure is the same kind of concept.

If we go up a level from here, we find the properties of substances on another level. For example, "index of refraction," which is how much light bends when it passes through something, or "surface tension," which is how water holds itself together. Both are expressed in numbers. This is due to the attraction of atoms etc. I remind you that it is necessary to scan many laws to see the source. But we still use the term "surface tension" and don't always care much about what's going on inside when we talk about it.

Let's go up a level in the hierarchy. When we look at the subject of water, we come across something called waves and storms. The word "storm" also refers to a very large collection of events. Then there are the "sunspots" which are "stars" which are clusters of objects. It's not always worth going back too far and thinking. We really can't. Because the higher you climb, the more steps are added weaker and thinner. enters. We haven't been able to look at them all at once.

As we move up this complexity ranking, we encounter things like muscle twitching or nerve irritation, which is something extremely complex in the physical world that requires arranging matter with extremely subtle complexity. Then things like "frogs" come along. We continue to meet; "people", "history", "politics", etc. we come to words and concepts, a set of concepts that we use to understand things at a higher level; As we keep ascending, we reach things like evil, beauty, hope.

If we use a religious metaphor, which end is closer to God? Beauty and hope or basic laws? I think what needs to be said is this: We need to look at all of the intertwined connections of being. All sciences, not just sciences but all intellectual endeavors, seek connections between hierarchical levels above and below; Beauty with history, history with human psychology, human psychology with brain functions, brain with neural impulses, neural impulses with chemistry, etc. Efforts to make connections We don't do that today. It's no use fooling yourself into believing that we can draw a straight line from one end of this thing to the other; 'Cause we're just beginning to realize

I don't think either extreme is closer to God. It is wrong to stand at either end, to walk only at that end of the pier, and to believe that a full understanding of what is going on will happen in that direction. Standing on the side of evil, beauty and hope or fundamental law; Everyone

It is wrong to hope that the only way to grasp the world will be through depth. It is not reasonable that those who specialize in one end ignore those who specialize in the other. The great crowd that works between these two extremes constantly connects one step to the other and allows us to understand the world better and better. In this way, working both at the ends and in the middle, we begin to gradually understand the extraordinarily large world of this nested hierarchy.

Probability and uncertainty nature related to quantum mechanics

In the past stages of the experimental observation process or the scientific observation of anything, it was intuition that provided a plausible explanation for events. Intuition comes from our simple experiences with everyday things. If we try to explain

more comprehensively and coherently what we see, as the field expands and we encounter more diverse phenomena, explanations become what we call laws and not simple explanations. Laws have a special quality; they seem to be moving further and further away from common sense and the intuitively obvious. Let's take the theory of relativity as an example. The suggestion is: if you think two things are happening at the same time, it's your blood; someone else decides
can remove; therefore the status "at the same time" is only a subjective impression.

There is no reason to expect otherwise. Because everyday life experiences are about things that move very slowly or about very specific circumstances; therefore they represent very limited phenomena in nature. A very small part of natural phenomena can be understood through direct experience. Only through accurate measurement and careful experimentation do we gain a broader perspective. Then we see unexpected things; very different than we can imagine, more than we can imagine... Our imaginations are stretched to the limit;

Understanding things that exist, not imagining things that don't exist, like in fictional novels. I want to talk about that.

Let's start with the history of light theory. It used to be thought that light behaved like rain, like bullets fired from a gun, like a shower of particles, particles. Further research revealed that this was not true, and that light actually behaved like a wave, like ripples in water. Then, in the 20th century, new research gave me the impression that light really does behave like particles in many ways. These particles could be counted by photoelectric effects - now they are called photons. When electrons were first discovered, they behaved just like particles, like spheres. More experiments; For example, electron diffraction experiments have shown that electrons behave like waves. Confusion grew over time about how electrons behaved - wave or particle, particle or wave? The available data showed that they were both similar.

This growing confusion was resolved in 1925 or 1926 when the correct equations for quantum mechanics were found. We now know how electrons and light behave. How do they behave? I would give the wrong impression if I said that they behave like particles. When I say they behave like waves, it's the same thing. They act in a unique, one-of-a-kind way. Technically we could call it a "quantum mechanical behavior". This is behavior like you've never seen before. Your experience of what you have seen before is not complete. As for how things behave on a very small scale, all we can say is that they behave differently. An atom does not behave like a swinging weight, suspended from a spring end. Nor does it behave like a miniature solar system with tiny planets moving in orbits. It also doesn't look like a layer of cloud or fog surrounding the core. It behaves like nothing you've seen before.

We can at least make one simplification: electrons behave in a way exactly like photons; both are "weird" but in the same way.

It takes a lot of imagination to see how

they behave; because what we are going to explain is different from anything you know. At least in this aspect, insofar as it is abstract and different from our experience, this course will perhaps be the most difficult course in the series. There's no way I can prevent that. I would remain incomplete if I did not give a series of lectures on the properties of the physical laws, and did not talk on a small scale about the actual behavior of particles. What I'm going to talk about is a universal feature that is unique to all particles in nature.

So, if you want to know the properties of the physical laws, this particular topic needs to be explained.

This is going to be difficult. But in reality this difficulty is psychological, one keeps asking oneself "but how can this be?" It stems from the boredom of asking. This question you are asking is an expression of an irresistible, but utterly impossible, desire to see her as something familiar. I won't explain it by analogy with the usual; I'll just explain. The newspapers once said that only twelve people understood the theory of relativity. I don't think there was ever a time like this. There may have been a time when only one person understood him; because he was the one who made this theory a reality even before I wrote it. However, many who read his work understood the theory of relativity in one way or another; their number was undoubtedly more than twelve. However, I can safely say that nobody understands quantum mechanics. So don't take the lesson too seriously and think you really need to understand what I'm going to say; Relax and enjoy! I will tell you how nature behaves. You will find him very sweet and charming if you accept that he can act like that. If you can, you constantly ask yourself "but how can this be?" don't ask; for your effort is in vain; You end up in a dead end from which no one has escaped yet. Nobody knows why that could be. for your effort is in vain; You end up in a dead end from which no one has escaped yet. Nobody knows why that could be. for your effort is in vain; You end up in a dead end from which no one has yet escaped. Nobody knows why that could be.

Now I'll explain to you how electrons and photons behave quantum mechanically,

using both analogies and opposites. If we only start from analogies, we will not succeed; the things we are going to explain with similar and contradictory aspects to what we know.
need to be treated. I do the analogy and contrast using first projectiles for particle behavior and then water waves for wave behavior. I will arrange a special experiment for this. Let me tell you how the experiment would play out if I used particles first; what could then happen to the waves; Finally, I will describe what happens when there are actually electrons and photons in the system. I will expose you to the quirks, mysteries and paradoxes of nature in this one experiment that contains all the mysteries of quantum mechanics. Do you remember the "two-hole experiment" or other situations in quantum mechanics? It's the same". Now I'll tell you about the two-hole experiment. The experiment contains all this incomprehensible. I won't skip anything and I'll reveal to you nature in its most elegant and difficult form, in all its nudity. We'll start with the shells first (Figure 28).

Figure 28

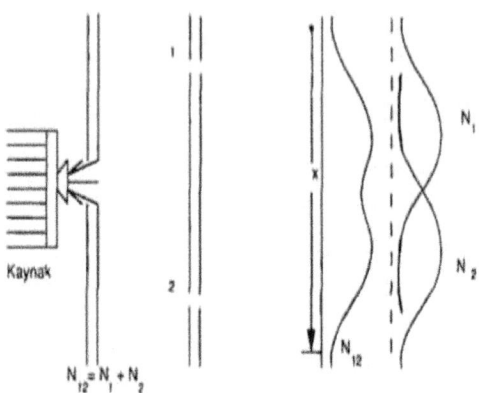

Suppose we have a machine gun as a bullet source, a plate with a hole in front of it with a hole for bullets to pass through, and the plate is bulletproof. Let there be a second plate further along with two holes - the famous two holes. I call them hole 1 and hole 2 as I will refer to these holes often. You should think of the holes as round holes in three dimensions; The picture only shows the cross section. Again, after a long pause, a curtain was raised. Detectors can be placed at different positions on the obstacle; this can be, for example, a box filled with sand that enables balls to be found and counted. In the experiments that I will conduct, I'm going to put this detector or box in different positions and count how many bullets fell in. In doing so I measure the distance of the box from a certain point and call it "x" and when we change the "x", so I'm going to talk about what happens when we move the detector box up and down. First I will make some changes in real shells with three abstractions. According to the first, since the machine gun is shaky and wobbly, the bullets go in different directions, and not straight, and can ricochet off the sides of the hole in the bulletproof panel. Secondly, we will say that although it is not so important, the speed or energy of all bullets is the same. With real spheres, the most important abstraction is that these bullets are absolutely indestructible. We want to find whole bullets in the box, not chunks of lead or half bullets. Think indestructible shells and soft armor. According to the first, since the machine gun is shaky and wobbly, the bullets go in different directions, and not straight, and can ricochet off the sides of the hole in the bulletproof panel. Secondly, we will say that although it is not so important, the speed or energy of all bullets is the same. With real bullets, the most important abstraction is that these bullets are absolutely indestructible. We want to find whole bullets in the box, not chunks of lead or half bullets. Think indestructible shells and soft armor. As the machine gun is wobbly and wobbly, according to the first, the bullets go in different directions, and not straight, and can ricochet off the sides of the hole in the bulletproof panel. Secondly, we will say that although it is not so important, the speed or energy of all bullets is the same. With real bullets, the most important abstraction is that these bullets are absolutely indestructible. We want to find whole bullets in the box, not chunks of lead or half bullets. Think indestructible shells and soft armor. The main abstraction is that these shells are absolutely indestructible. We want to find whole bullets in the box, not chunks of lead or half bullets. Think indestructible shells and soft armor. The main abstraction is that these shells are absolutely indestructible. We want to find whole balls in the box, no pieces of lead or half bullets. Think indestructible shells and soft armor.

The first thing that strikes us about

these clams is that they are all in flakes. The incoming energy is a sphere: a single bang! If you count the balls, there are one, two, three, four balls and they came one after the other. all the same size We assume that. Everyone is either all-in or all-out when in the box. If we set up two separate boxes and assume the rifle isn't fired very quickly, there will be plenty of time between shots to observe them. Two grenades will not enter the boxes at the same time. You slow down the shot of the rifle and take a quick look at the two boxes, you can't see that two bullets go into both boxes at the same time; because each sphere is a single specific mass.

Now I want to measure how many rounds arrive in a given period on average. Let's say we wait an hour, count how many balls are in the sand and take the average. We can say that the number of bullets arriving in an hour is "probability". Because this number gives us the chance that a ball will land through the hole in the box. Of course, if I change the "x", the number of balls in the box also changes. In the figure I plotted the curve I obtained when holding the box in each position for an hour and plotting horizontally the number of balls in the box. In this case I get a curve that looks something like the N12 curve. Because if the box is behind one of the holes, a lot of balls will fall in, if it is in a slightly different state fewer balls will fall in as the balls will bounce off the sides of the hole and as a result, it will curve, drop to zero at the ends. The corner looks like N12, I call the number N12 when both holes are open; this is the number of balls entering the box from both holes 1 and 2.

I want to remind you that the number I'm graphing is not made up of whole numbers. This can be any number. Although the bullets come whole, it can be two and a half bullets an hour. Two and a half laps per hour means twenty-five laps if you run it for ten hours, so an average of two laps per hour

means half a sphere. I think you all know the jokes about the average family in the US with two and a half kids. This does not mean that there is only half a child in every family; Children come whole! However, if we take the average per family, the result can be any number. N12 is also one such number;
indicates the average number of laps in the box per hour and does not have to be an integer. What we measure is the "probability of occurrence", a technical term for "the average number of arrivals in a given period of time".

If we examine the N12 curve, we can easily interpret it as the sum of two curves: the sum of N1, the number of balls, when hole 2 is covered by another plate placed in front of it, and N2 when hole 1 is closed. We now discover the very important law: if both holes are open, the number obtained is equal to the number of hole 1 plus the number of hole 2. The important thing about the theorem that says we just have to add these two is that absence of interference.

N12 = N1 + N2 (no interference)

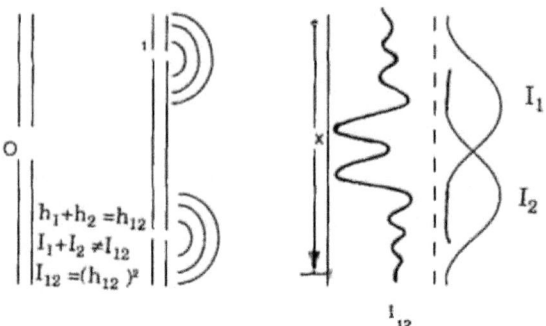

Figure 29

We found this result for spheres; That part of the job is done. Now let's repeat what we did for the spheres with the water waves (Figure 29). In this case, the source is a large mass that we shake up and down in the water. A lead-resistant slab is also a barrier from breakwaters or barges with a gap for water to pass through. It would be wiser to do this experiment with small waves instead of big ocean waves. I wave my finger up and down and use a piece of wood with a hole for the waves to pass through as a barrier. Then I added a second barrier with two holes and finally a detector. The detector detects how much the water is vibrating. For example, I put a cork in the water and measure its up and down movements. What

I'm going to measure is actually the energy of the cork vibrations; This is exactly proportional to the energy transported by the waves. I'm assuming the ripple is done smoothly and perfectly, so the ripples are evenly spaced. One important thing about water waves is that what we measure can be any size. We measure the intensity of the wave or the energy of the mushroom. If the waves are very calm and I shake my finger a little, the movement of the cork will also be small. Energy, no matter how much, is proportional. It can be any size; does not come in whole; it's not all or nothing.

What we are going to measure is the intensity of the waves; or more specifically, the energy that waves produce at a point. When we measure this intensity, what do we see? I call it I to emphasize that it's not just about the number of particles, it's about the intensity. The curve with both holes open is shown in the figure (Figure 29). This is an interesting looking curve. If we move the detector to different positions, we get an intensity curve that changes very quickly and strangely. maybe you know why The reason is this: as the waves progress, ridges and troughs emanate from both hole 1 and hole 2. At a point equidistant from both holes, since they arrive there at the same time, the two waves overlap and the ripple increases. In the middle there are many vibrations. On the other hand, if the detector is from hole 2, if I take it to a point further away from hole 1, it will take longer for the waves to come from hole 2 than if they come from hole 1; When a peak from hole 1 reaches the detector, the peak from hole 2 has not yet been reached; rather it has reached a bottom of hole 2. Therefore, under the action of waves, the water tries to rise from the two holes, tries to sink and, as a result, remains motionless; or almost immobile. Therefore, a slight bump is formed at this point. Further there is enough delay and the two wave crests overlap and a large crest arises again, although one is a whole wave behind the other. So depending on the "interference" of the ups and downs, then large When a peak from hole 1 reaches the detector, the peak from hole 2 has not yet been reached; rather it has reached a bottom of hole 2. Therefore, under the action of waves, the water tries to rise from the two holes, tries to sink and, as a result, remains motionless; or almost immobile. Therefore, a slight bump is formed at this point. Further there is enough delay and the two wave crests overlap and a large crest arises again, although one is a whole wave behind the other. So depending on the "interference" of the highs and lows, then large If a peak from hole 1 reaches the detector, the peak from hole 2 has not yet reached; rather it has reached a bottom of hole 2. Therefore, under the action of waves, the water tries to rise from the two holes, tries to sink and, as a result, remains motionless; or almost immobile. Therefore, a slight bump is formed at this point. Further there is enough delay and the two wave crests overlap and a large crest arises again, although one is a whole wave behind the other. So depending on the "interference" of the ups and downs, then big tries to descend and as a result remains motionless; or almost immobile. Therefore, a slight bump is formed at this point. Further there is enough delay and the two wave crests overlap and a large crest arises again, although one is a whole wave behind the other. So depending on the "interference" of the ups and downs, then large tries to dismount and remains motionless as a result; or almost immobile. Therefore, a slight bump is formed at this point. Further there is enough delay and the two wave crests overlap and a large crest arises again, although one is a whole wave behind the other. So depending on the "interference" of the ups and downs, then big small, then large, small again... The word company is used differently here, specific to science. We can get what we call "constructive interference," the type of interference where two waves interfere and increase in intensity. The point is that I12 is not equal to I1 plus I2; we say that constructive and destructive initiatives take place. To find the shape of I1 and I2, we draw I1 by closing hole 2 and draw I2 by closing hole 1. The violence we find when we close a hole is the violence that comes out of a single hole without interference. These curves are shown in Figure 29. As we can see, 11 is similar to both I2 and N2; but completely different from N12.

 Actually the math of the I12 curve is quite interesting, it is true that the water height which we will call h is equal to the sum of the water heights when hole 1 is open and when hole 2 is open. Since the height of hole 2 is negative for the dimple point, what came from hole 1 is thus removed. You can express this in terms of water height.

But in any case, for example if both holes are open,

It can be seen that the intensity is not equal to the height, but is proportional to the square of the height. Since the squares are occupied, we find these curves very interesting.

$h12 = h1 + h2$ but

$I12 \neq I1 + I2$ (fault)

$I_{12} = (h_{12})^2 \, I2 = (h2)2$

This is the case with water. We now consider the same for electrons (Figure 30).

Figure 30

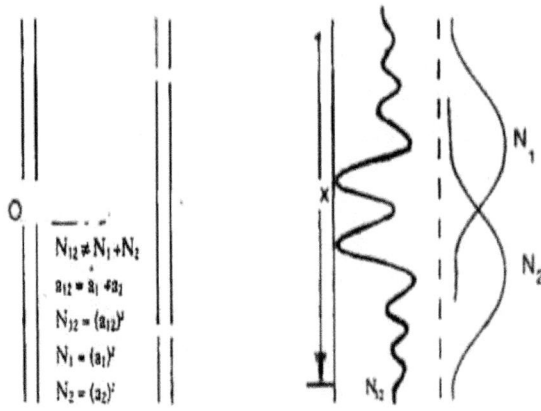

A filament as a source; a tungsten sheet with holes as a barrier; And as detectors we will use any electrical system that can detect the charge of an incoming electron with sufficient accuracy by carrying the energy of the source. If we want we can use photons instead of electrons, black paper instead of tungsten sheet - we have to find something better as the fibers in black paper are not suitable for making sharp-edged holes - and as a detector we can use a photomultiplier the incident photons can distinguish individually. What happens in both cases? I will describe the experiment with electrons only; The situation is similar with photons.

First, a crackling sound can be heard from the detector, behind which is a sufficiently powerful amplifier. The crunch is of definite and constant intensity. If you attenuate the source, the crackling sounds are heard with the same intensity but less frequently. As you crank up the source, the crackle speeds up and clogs the amp. The source you are using as a detector You should turn it down so it doesn't get the popping that exceeds the device's capacity. If you then place a detector in a different spot and listen to both at the same time, you won't hear two crackles; at least if the source is weak enough and the time measurement is sensitive enough. If the intensity of the source is reduced so that the electrons arrive less and less, there is no simultaneous cracking from the two detectors. This means that what comes in is in the form of particles (particles of a certain size, each going to one place). All right, so electrons and photons come as particles. So we can repeat exactly what we did for the balls and measure the probability of them coming. What we do is hold the detector at different points, say for an hour to measure how many electrons arrive at the end of each hour and take the average. What do we find for the number of incident electrons? The kind we found for bullets? What happens for N12 in Figure 30, that is, when both holes are open?

It is shown that we have found. This is the behavior of nature; we have found the same curve for the interference of waves. For which situation was this curve obtained? It is not given for the energy of the wave but for the probability of it occurring in particles.

The math for this is simple. We replace I with N, and we need to replace h with something new since it doesn't indicate the height of anything. We invent an "a" and call it probability amplitude; because we don't know what it is. Then a1 is the amplitude of the probability of coming out of hole 1; a2 represents the amplitude of the probability of coming out of hole 2. Come To find the amplitude of the probability, add the two and square them.

we get. Here we make a direct analogy to

what we did for waves. Because we need to get the same curve; We use the same math.

But it is necessary to check one point, the initiative. I didn't mention what happens if we close one of the holes. Now let's look at this interesting curve obtained when electrons come out of only one of the holes. We close one of the holes and find out how many passes go through hole 1 and get the simple N1 curve. Close the other hole If we measure it, we get the N2 curve. But when these two are added together, we don't get the curve for N1 + N2; company gets involved. In fact, the math for this is different. The probability of occurrence is a square of the amplitude, and the amplitude itself is the sum of two things: $N_{12} = (a_1+a_2)^2$. The problem here is: while the electron is scattered in one direction when passing through hole 1 and in the other direction when passing through hole 2, why isn't it the sum of the two if both holes are open? For example, if I put the detector at point q with both holes open, almost nothing comes out; but when I close one hole many things come, and when I close both holes some things come. I leave both holes open, nothing comes out; If I let them come through both holes, they still won't come. Let's take the center. We can show that this is slightly more than the sum of the curves of the two individual holes. If I were smart enough they would somehow go back and forth through the holes, or something complicated.

You might think I could come up with something to explain this event, like what they did or that some of them split in half and went through both holes. However, since both curves and mathematics are very simple (Figure 30), no one has yet been able to provide a satisfactory explanation. In summary, electrons are like particles; However, the probability of these particles coming is determined as in the wave intensity, in this sense electrons are sometimes called waves and sometimes behaves like a particle. They behave in two different ways at the same time (Figure 31).

Figure 31

That's all there is to say. In principle, if I gave a mathematical expression to find the probability of arrival of electrons in each of the states, this would be over. However, there are some subtleties in the fact that nature behaves this way and I want to discuss them; because they may not be obvious at this stage.

First, let's consider a proposition that we might find plausible since these things are particles. Since incoming things, namely electrons, are always in the form of a whole particle, it is reasonable to assume that an electron will go through either hole 1 or hole 2. If it's a particle, it's obvious that nothing else can. Since I am now going to discuss this phrase, it is necessary to give it a name; I call it suggestion A.

<u>Suggestion A:</u>
The electron either passes through hole 1 or hole 2.

We've already explained a bit about the outcome of Set A. If it's true that an electron went through either hole 1 or hole 2, that

would be the total number of electrons that came in.
must be distinguished as the sum of the two. The total is the sum of those arriving via hole 1 compared to those arriving via hole 2. However, since the overall curve doesn't differentiate into such a nice sum of the two curve segments, and since experiments determine the number of incoming electrons when one or the other of the holes is open, we must conclude that Theorem A must be wrong, since Experiments that determine the number of incident electrons do not conclude that the total number is the sum of the two. If it's not true that electrons only go through hole 1 or hole 2, maybe they temporarily split in two or something like that. So statement A is false. That's logic. It's maybe regrettable, maybe happy

We just have to watch them. We need light for observation. So we put a very strong light source behind the holes. The light is scattered by the electrons and bounces off when it hits them. If the light is strong enough, electrons can be seen passing by. So if we stand behind hole 1 or hole 2 and count the electrons, we're looking for a flash of light behind hole 1 or hole 2, or a "half glow" in both at the same time before the electron is counted. We'll find out what's going on. We turn on the light and start looking; and voila: whenever there is a number in the detector, a glow will appear behind either hole 1 or hole 2. We see that the electron comes out of hole 1 or hole 2 exactly 100%. There is a paradox here!

Now we're going to corner nature a bit. I'll tell you what to do. We leave the light on and count how many electrons have passed. We will arrange two columns for hole 1 and hole 2 and mark each electron entering the detector in the appropriate column. What is the sum found for hole 1 at different positions of the detector? What do I see when I look behind hole 1? What I see is the N1 curve (Fig 30). The distribution of this column is the same as we expected when we closed hole 2; It's the same whether we look or not. Closing hole 2 we find a distribution as if we were looking through hole 1; The number of corners of hole 2 is also a simple N2 curve. Now the number of all arrivals must also be the total number, that is, the number plus the number N2; because each of the arrivals was marked in column 1 or column 2. The number of arrivals must be equal to the sum of these two, and the distribution must be N1 + N2, but I said it's curved. No, the distribution is N1 + N2. It really is; it should be so and it is so. If we display the results with the sign " " when the light is on, we see that N1 is the same as N1 when there is no light, and N2 is the same as N2. However, when the light and both holes are on, we see that N' 12 is equal to the sum of hole 1 and hole 2. This is the result when there is light. I find different results when I turn the light on or off. When I turn on the light, the distribution is N1+N2, and when I turn it off, it's N12. Let's turn the light back on; the result is again N1 + N2. As you can see, we cornered nature! So we can say that the light affects the result. The result found when the light is on differs from that found when the light is off. Light also affects the behavior of electrons. So we can say that the light affects the result. The result found when the light is on, differs from that found when the lights are off. Light also affects the behavior of electrons. So we can say that the light affects the result. The result found when the light is on differs from that found when the light is off. Light also affects the behavior of electrons.

you can tell. In this slightly flawed experiment, if you talk about the movement of the electrons, you could say that the light affects the movement. Electrons arriving at the maximum are deflected and repelled by the light and arrive at the minimum, flattening the curve to give the simple N1+N2 curve.

Electrons are very sensitive. Lighting a baseball doesn't change anything; Baseball continues in the same direction. But when you shine light on an electron, the electron gets slightly distorted and does something different than it normally does; because you turned on the light and the light is very strong. Now let's gradually dim the light until the surroundings are very dark and build in sensitive detectors that can see in the dim light and search in that light. As the light becomes weaker and weaker, the electrons of such weak light change from N12 to N1 + N2

So you wouldn't expect it to be 100%

impressive. As the light fades, the environment should become less and less light. So how does one curve become another curve? Light, of course, is not like a water wave. Light also has the property of a particle, which we call a photon. As you decrease the intensity of the light, you decrease the number of photons emitted by the source. The number of photons hitting an electron is at least one. Sometimes when there are very few photons, the electron may have passed in a state where the photon was not present; then you don't see it. That means a very faint light means fewer photons, not a very small effect. So for a very faint light I need to add a third column under the heading ".

are distributed accordingly. As I dim the light, what I see will gradually decrease and what I don't see will increase. In any case, the true curve is a mixture of the two curves. As the light decreases, the curve becomes more and more like N12.

It is impossible for me to discuss all the different methods we could propose to find out which hole the electron went through. However, it turns out that adjusting the light so that we can tell which hole the electrons are going through without affecting their arrival patterns and eliminating interference is impossible. Not just for light, but for everything else, whatever you're using, it's basically impossible. If you want, you can find many ways to determine which hole the electron went through. It turns out he's been through one or the other every time. However, if you want to prevent your device from affecting the movement of the electron, you will not always know which hole the electron went through,

When Heisenberg discovered the laws of quantum mechanics, he realized something: the laws he discovered were only consistent when some fundamental limitations were imposed on our ability to experiment that we hadn't realized before. In other words, you can't be as specific as you want with experiments. Heisenberg proposed an uncertainty principle. In the context of our experiment, we can put it like this (He put it differently. However, both are completely equivalent; you can switch from one to the other): "It is impossible to build a device that detects which hole the electron is goes through and at the same time does not affect the electron so much that it destroys the interference pattern. Nobody has found a way to prevent this. all from you I'm sure you're looking forward to discovering ways through which hole the electron went; However, if you examine each of them carefully, you will see that there is a bug. You might think you can do this without affecting the electron; But there will always be some error, and it will always turn out that the difference in the curves is caused by devices that detect which hole the electron goes through.

This is a fundamental feature of nature and tells us something that applies to everything. If tomorrow a new particle called "kaon" is discovered - indeed the kaon has been discovered; I only used it to give it a name - and if I use its interaction with the electron to find out which hole the electron went through, I think I already know something about the behavior of this new particle: whether I can tell which hole the electron went through, its work being done without affecting the electron by an interference pattern that the interference can't do without turning it into a non-order. Therefore, the uncertainty principle can be used as a general principle that allows us to predict many properties of unknown objects.

Let's go back to theorem A. "Electrons have to go through one hole or the other".

Is that right or wrong? Physicists have a few methods to avoid falling into the trap. They explain their rule of thought like this: if you have a device (and you can have such a device) that can tell which hole an electron goes through, then you can say that it either goes through one hole or the other; it is possible. It always goes through one or the other hole if you look. However, if you don't have a device to detect which hole it goes through, it.

(You can always say that as long as you stop thinking immediately and don't draw any conclusions from it. Physicists prefer not to think about silence now). It would be a wrong prediction to say that the electron goes through this hole or the other when you're not looking. If we want to interpret nature, this is the logical tightrope on which we will walk like acrobats.

The sentence I am talking about is not a

general sentence. It's a proposition, not just for two holes, that can be expressed as follows. In an ideal experiment - that is, an experiment in which everything is determined as accurately as possible - the probability of an event occurring is the square of something. In our example I'll call that "a" which is the square of the amplitude. If a phenomenon can occur in different ways, the amplitude probability, i.e. this number "a", is the sum of the "a" found for each option. When an experiment is conducted to determine which option is used, the probability of the event occurring will be different; is the sum of the probabilities of each option. So you lose the attempt.

Our question now is how this actually happens. What is the mechanism that causes this? Nobody knows any mechanism. Nobody can give you a deeper explanation of this phenomenon than what I have told you. There may be those who give more detailed explanations by doing other experiments showing that it is impossible to determine which hole the electron goes through without removing the interference,

and those who come from experiments other than interference experiments with speak two holes. However, these are only repetitions for better understanding. They're not deeper, they're wider. The mathematical expression can be given in more precise forms; one can say that they are not real but complex numbers;
or a few other points unrelated to the main idea. The deep mystery, however, is nothing other than what I am describing; Nobody has dug deeper yet.

So far we have calculated the probability that an electron will come. The question now is whether there is a way to find out where a particular electron really comes from. We are not opposed to using probability theory, i.e. calculating probabilities when the situation gets too complicated. We throw dice in the air; We admit that we don't know enough details to make a safe prediction given various resistances, atoms and all that complex stuff; so we calculate the probability that it will happen either way. What we are proposing here is that deep down there is a probability that unexpected things will happen in the fundamental laws of physics.

Suppose I set up an experiment where interference occurs when the light is turned off. Then I say that even with light I cannot know which hole the electron went through. All I know is that I'm either looking at this hole or the other. There's no way I can predict which hole I'm going through. In short, the future is unpredictable. Using all the information available, it is impossible to know in any way which hole the electron will pass through or in which hole it will appear. One consequence of this is that if the physicist's original goal was, as is commonly believed, to have enough information to predict what will happen next under certain conditions, he seems to have given up. Here are the conditions: Electron source, strong light source, tungsten sheet with two holes. Now can you tell me which hole I will see behind the electron? According to one theory, the reason it is not known which hole the electron will go through is because before is that it's predetermined by some complicated things in the source, there are internal gears, internal gears that decide which hole it goes through. The probability is half; it falls randomly just like a die. The physics are not yet complete; If it is sufficiently complete, we can predict which hole it will go through. This is called "hidden variable theory". This theory cannot be true; Our reason for not being able to make predictions is not due to a lack of detailed information.

  I said if I don't turn on the light, I get the initiative. If there is a condition where I get this interference phenomenon, it is impossible to consider it in relation to going through hole 1 and hole 2; because this interference curve is very simple and mathematically very different due to the contribution of the other two probability curves. If we could tell which hole an electron would go through if the light was on, it wouldn't matter if the light was on or not. We could observe, in the absence of light, the cogs that exist in the source that we see and that allow us to say that it will go through hole 1 or hole 2; So we could tell which hole the electron went through when there was no light. If we could do this, the curve, that we would get expressed as the sum of what went through hole 1 and hole 2; but it can't. So, whether the light is on or not, in any case where the experiment is set up to create interference without light, it should be impossible to know in advance which hole the electron will pass through. It is not because of our ignorance of the inner cogs, the inner turmoil, that nature seems to have possibilities in her structure. It's like something that exists in the inner structure of nature. Someone put it this way: "Not even nature itself knows which direction the electron will go." in which the experiment is arranged to produce interference in the absence of light, it should be impossible to know in advance through which hole the electron will pass. It is not because of our ignorance of the inner cogs, the inner turmoil, that nature seems to have possibilities in her structure. It's like something that exists in the inner structure of nature. Someone put it this way: "Not even nature itself knows which direction the electron will go." in which the experiment is arranged to produce interference in the absence of light, it should be impossible to know in advance through which hole the electron will pass. It is not because of our ignorance of the inner cogs, the inner turmoil, that nature seems to have possibilities in her structure. It's like something that exists in the inner structure of nature. Someone put it this way: "Not even nature itself knows which direction the electron will go." that exists in the inner structure of nature. Someone put it this way: "Not even nature itself knows which direction the electron will go." that exists in the inner structure of nature. Someone put it this way: "Not even nature itself knows which direction the electron will go."

    A philosopher once said, "For science to exist, similar conditions must lead to similar results." Fine, but they don't lead. You determine the situation under the same conditions every time and cannot predict which hole behind the electron you will see. But even if similar conditions don't always lead to similar results, the science endures. Not knowing exactly what will happen in advance makes us miserable. There can be very dangerous and serious situations to be aware of; but you still can't know them in advance. For example, we can build a photoelectric cell and an array through which a single electron can pass - we shouldn't do it, but we can. If we see the electron behind hole 1, let's activate the nuclear bomb and go to III. We can start World War II. But if we see him past hole 2, the peacetime antennae will come out and delay the war for a while. No matter how far science advances, man's future will depend on something he cannot predict. The future is unpredictable.

What is necessary for the "existence of science" is not determined by natural properties, by presuppositions of splendour; they are always determined by the matter we are working with, by nature itself. We will search and see what we find. But we cannot predict exactly what will happen. The most conceivable possibilities often do not turn out to be true. If science is to advance, you have to experiment, explain the results honestly—the results have to be explained before anyone tells you how you want them—and finally have the intelligence to interpret the results. An important point with this intelligence is not to be too sure beforehand what the results should be. They can be biased and say "this can't be it can be tasted. Being biased is different from being absolutely sure. Definitely not biased; I just mean the trend. It doesn't matter if you're just inclined; because if your inclination is wrong, the successive test results will constantly annoy you, and in the end you can no longer ignore them. However, if you are very sure of what the pre-science prerequisite must be, you can ignore the results. Indeed, for science to exist, we need brains that do not impose preconditions on nature like our philosopher's.

The search for new laws

The subject I want to talk about in this lecture is not exactly the nature of the laws of physics. When we talk about the nature of the physical laws, it is assumed that we are at least talking about nature. But I don't want to talk about nature, I want to talk about where we are relative to nature. I want to talk to you about what we think we know, what predictable things are, and how those predictions are made. It has been suggested to me that the ideal method is to explain step by step how a law is predicted and then discover a new law for you. I don't know if I can.

First I will tell you what the current situation is, what we know about physics. During these lessons you already know how You might think I told you everything; for I have told you all known great principles. But principles must be principles about something. Conservation of energy is about the energy of something; the laws of quantum mechanics are quantum laws about something; but all these principles taken together say nothing about the content of nature of which we speak. I'm going to talk to you now about what all these principles are supposed to apply to.

First, there is a "thing" called matter, and it's interesting that all matter is the same. It is well known that the matter that makes up the stars is the same as the earth is made of. The quality of the light emitted by these stars is a kind of fingerprint, which tells us that there are also atoms similar to those on Earth. It can be seen that in living things there are atoms of the same kind as in inanimate ones. Frogs are made up of the same "set" of atoms as rocks; but the rules are different. This makes our problem easier; there are only similar atoms everywhere.

The atoms all appear to have the same general structure. They have a nucleus and electrons around the nucleus. We can make a list of the parts of the earth that we assume we know (Figure 32).

Figure 32

First of all, in the outer part of the atom there are particles that we call electrons. Then the kernels; Today, however, it is known that these too consist of two different particles called neutrons and protons. You see stars, you see atoms and they emit light. Light itself is described by particles called photons. We talked about gravity in the beginning; If quantum theory is correct, there must be some kind of gravitational wave that behaves like a particle; We call these particles gravitons. If you don't believe it, you can call it gravity and pass it on. Finally, I talked about something we call beta decay. Here the neutrons dissociate into a proton, an electron and a neutrino (more precisely an antineutrino, because there is another particle called neutrino). All the particles I have included in the list also have antiparticles. With this brief explanation we avoid doubling the number of particles; there is no confusion there either.

As far as we know, low-energy phenomena occurring everywhere in the universe, i.e. all normal phenomena, can be explained by these particles that I have listed. There are some exceptions here and there caused by high energy particles, we also managed to do some "weird" things in the lab. Apart from these special cases, all events known to us can be explained by the effects and movements of these particles. For example, it is accepted in principle that life itself can be explained in terms of the movements of atoms; These atoms are made up of neutrons, protons and electrons. I should add right away that by "in principle" I mean this: if we can understand everything, we don't think there's anything new in physics that we need to discover, to understand the phenomenon of life. Another example is when stars emit energy (stellar or solar energy).

It is the assumption that particles can be explained by nuclear reactions. According to the current state of knowledge, this atomic model can be used to precisely explain every detail of the behavior of atoms. I can even say: I think that of all the facts that we know today, there is not an event that we are sure cannot be explained in this way or that contains a deep mystery.

This was not always possible. For example, there is a phenomenon called superconductivity, which means that metals conduct electricity without resistance at low temperatures. It was not previously known that this was the result of known laws. In the meantime, however, it has become clear that, with careful consideration, this can be fully explained with our current knowledge. There are other phenomena that cannot be explained with our physical knowledge, such as extrasensory perception (sixth sense), the existence of which we are not entirely certain. However, if their existence can be proven, then the physics is proved not to be complete; Therefore, their existence is very important for physicists. There have been many experiments that have shown that they do not exist. Astrological influences are also in the same situation. If it is true that the stars can influence the most favorable day to visit the dentist - in America there is such a use of astrology - then the theory of physics will be proved wrong. Because in principle there is no known mechanism that provides this with the behavior of particles. Therefore, there is distrust of such ideas among scientists.

However, in the early days when hypnosis was not fully understood, it was also thought to be impossible. It is now better understood that hypnosis is the result of a physiological process that is not yet known but is normal.
it was recognized that it is not impossible; It seems that it does not require any special force.

Today, our theories are sufficiently complete and precise to explain what goes on outside the nucleus. That is, given enough time, anything can be calculated with as much accuracy as it can be measured. However, the forces between the neutrons and protons that make up the nucleus are not as well known and not well understood. I mean it in this sense: today (even if we have enough time and computers) we don't understand the forces between neutrons and protons well enough to calculate the energy levels of carbons or the like; What we know is not enough. Although we can do these calculations for the energy levels of electrons outside the atom, we cannot do them for the nucleus because nuclear forces are not yet well understood.

To learn more about them, experimenters studied them at very high energies, colliding protons and neutrons at very high energies and getting some strange results. We believe that by studying them, we will better understand the forces between the neutron and proton. With these experiments Pandora[26]The box has been opened! What we really wanted was just to learn more about the forces between neutrons and protons. However, when these particles collided, we discovered that there were other particles in the world. More than four dozen new particles have also emerged in an attempt to understand these forces. We add these four dozen different ones to the neutron/proton column (Figure 33); because they also interact with neutrons and protons and are also related to the forces between them. Also, we found two other things unrelated to nuclear forces when we scanned this huge swamp.

Figure 33

One is the mu meson or muon, the other the accompanying neutrino. There are two types of neutrinos, one with the electron and the other with the mu meson. It is very interesting that all the laws regarding the muon and its neutrino have been found. As far as we can understand experimentally, the law is as follows: the muon and its neutrino behave exactly like the electron and its neutrino; The only difference is that the mu meson weighs 207 times that of the electron. It's a bit strange that this is the only difference between these two objects. Four dozen new particles and their antiparticles form a frightening spike. These have different names; Mesons, pions, kaons, lambda, sigma... However, it's a bit easier for these particles to form groups. In fact, some of these particles are so short-lived that there is debate as to whether they can actually be counted. I will not interfere in these discussions.

To illustrate this grouping problem I will consider the neutron and the proton. The masses of the neutron and proton are equal, about one tenth of a percent. One weighs 1,836 times the electron, the other 1,839 times. What's even more interesting about nuclear forces is that the force between two protons is something like a proton with a proton in terms of strong forces in the nucleus.
be equal to the force between the neutrons; and the force between a neutron and a neutron is also the same. In other words, you cannot distinguish between a proton and a neutron in terms of strong intranuclear forces. This becomes a law of symmetry. Replacing protons with neutrons makes no difference (when it comes to strong forces). But when you replace a neutron with a proton, there's a big difference; Because the proton is electrically charged, the neutron has no electrical charge. With electrical

measurements, you can immediately see the difference between a proton and a neutron. Symmetry, in the sense of replacing one thing with another, is what we call "approximate symmetry". This applies to the strong interaction of nuclear forces,

As the particle groups multiply, shifting operations such as neutron replacement for protons can also be carried out between other particles. However, the level of accuracy is further reduced. The statement that protons can always replace neutrons is an approximation - not true for electricity. The shifts in this larger frame that we think are possible show an even weaker symmetry. However, these partial symmetries helped group particles, locating missing particles and finding new ones.

Such games, roughly assessing the relations between groups, are man's warming up with nature before he finds a really important and deep fundamental law. There are very important examples of this in the history of science.

For example Mendeleev's[27] Discovering the periodic table of the elements is similar to this game. This is the first step; A full explanation of the cause of the atomic table could come much later along with the atomic theory. Also what we know about intranuclear energy levels, Maria Mayer and Jensen[28] through what they called the shell model of the kernel. Physics is an analogy game aimed at reducing complexity through rough guesses.

In addition to these particles, all the principles we mentioned earlier; Symmetry, relativity, and the need for objects to behave according to quantum mechanics, and when you combine this with relativity, any conservation laws that occur must be local.

If we look at all these principles, we see that there are many and that they do not agree with each other. When we consider quantum mechanics, relativity, the statement that everything must be local, and some unspoken assumptions, the inconsistency arises. Because if we calculate different things, we get "infinity"; How can we say that it is compatible with nature when we get infinity? An example of the assumptions I mentioned that are not clearly articulated and whose true meaning we cannot grasp due to excessive bias is the following statement: If you calculate the probability for each possibility, e.g. B. 50% that this happens, 25% for this their sum must be 1. We think if you add up all the options, you should get a 100% probability. That makes sense. But difficulties always arise from the sensible. Another thesis is that the energy in something is always positive and not negative. Maybe another suggestion 'causality' is added before inconsistency appeared, ie the idea that effects cannot precede their causes. In reality, no one considers the proposition of probability or causation; Quantum mechanics, relativity, locality, etc. have not presented a model consistent with the principles. Hence we do not know from which assumption the difficulty of obtaining infinity arises. Nice problem! On the other hand, we know that with manual dexterity we can sweep infinities under the carpet and continue calculating for a while.

Well, that's the case now. I will now consider how I can search for a new law.

To find a new law, we usually use the following method: First we make a guess. Next, we calculate what conclusions could be drawn if the law we predicted is correct. Then, with the help of experiment or experience, we examine whether these conclusions are true in nature through direct observation. If they disagree with the experiments, they are wrong. This simple statement is the key to science. It doesn't matter if your guess is very good, you're very smart, who did it, or what their name is; the experiment is wrong if the prediction results contradict the experiment; that's it! A little more verification is needed to make sure it's wrong. Because the experimenter may have misrepresented the results or something was not considered in the experiment, e.g. B. Pollution; or, even if the person making the calculations is the same person making the estimation, he may have made a mistake in the estimation. Those are obvious things. That's why when I say it's wrong, if it contradicts the experiment, I'm saying that the experiment is verified, the calculations are verified, and the expected results are in fact the logical reason for what is being predicted. He may have made a mistake in the assessment. Those are obvious things. That's why when I say it's wrong, if it contradicts the experiment, I'm saying that the experiment is verified, the calculations are verified, and the expected results are in fact the logical reason for what is being predicted. He may have made a mistake in the assessment. Those are obvious things. That's why when I say it's wrong, if it contradicts the experiment, I'm saying that the experiment is verified, the calculations are verified, and the expected results are in fact the logical reason for what is being predicted.

I mean that it has been checked over and over again as a result of a carefully controlled experiment.

What I have told you may give you some misconceptions about science. Constantly making assumptions about possibilities and comparing them to experimentation can indicate that experimentation is in a somewhat inferior position. In fact, experimenters have a special quality; They like to experiment with things that no one has yet guessed. They often conduct experiments in areas where theorists have a reputation for failing to make predictions. For example, we have many laws; but we do not know if they are valid in the high-energy state. Whether they are valid in this case is just a guess. The experimenters performed experiments at high energy levels. Every now and then experiments cause some problems; That is, they show that something what we think is right is wrong. So an experiment can lead to unexpected results; which leads us to another guess. An example of such unexpected results is the mu meson and its neutrino. Nobody had suspected about its existence before it was discovered. To date, there is no estimation method to determine what the natural consequence of this is.

As you can see, we can consider proving any particular theory wrong using this method. If we have a particular theory, a true guess, and can draw conclusions from it that can be evaluated through experimentation, then in principle we have the ability to test all theories. There is always a chance to prove a particular theory wrong; But be careful, we can never prove it's true. Well, you guess the results Suppose you do calculations and the results you calculate agree with the experiment every time. So is the theory correct? No; it just hasn't been proven wrong. If you calculate larger results later, larger experimentation may be required and you may find that it is wrong. This is why Newton's laws of planetary motion have lasted for so long. Newton "guessed" the law of gravitation, calculated many results in this system, compared them with experiments. It took centuries for a small flaw in the movement of the planet Mercury to be uncovered. During all this time, the theory could not be proven wrong, so it was accepted as true for the time being. However, their accuracy has never been proven. For tomorrow's experiments may prove false what we believe to be true. We can never be absolutely right; only we can be absolutely sure of our mistake. But it's remarkable that we can still have ideas that endure for so long.

One way to thwart science is to conduct experiments only in areas where we know the laws are correct. However, empiricists put the most effort and work in those areas where the expectation that the theories are wrong is strongest. In other words, we try to prove our own wrong as quickly as possible. Because only then is progress possible. For example, nowadays we don't know where to look for the fault for normal low-energy phenomena, we think that everything is in place. Therefore, there are no special large-scale projects to find faults in nuclear interaction and superconductivity. In these lectures, I focus heavily on discovering fundamental laws. Superconductivity and Nuclear The understanding of interaction phenomena on another level, through fundamental laws, is also one of the interesting topics in physics. But the issue I am concentrating on now is the detection of flaws, the exposure of flaws in basic laws. All present-day experiments to discover new laws are high-energy experiments, since nobody in the field of low-energy phenomena knows what to study.

Another point I should draw your attention to is that a theory which is not clear and distinct cannot be proved wrong. If the prediction you made is not well expressed and far from clear, the method you use to find the results is also not clear and clear. You are insecure and say: "I think everything is right; because this and that or this and that, it's more or less like that; I can explain a bit how this works..." You think that's a good theory; because it is not demonstrably wrong. Although the process of calculating the results is uncertain, with a little skill any experimental result can be converted into something similar to the expected result. You may have encountered something similar in other areas. "A" hates his mother. The reason for this is of course The reason is that his mother did not love and caress him enough as a child. If you search, you will find that his mother loved him very much and everything was going well. So the reason is that her mother spoiled her too much as a child! With an ambiguous theory, it is possible to arrive at both conclusions. The remedy is this: If it were possible to determine in advance exactly how much love was inappropriate and how much love was excessive, then there would be a valid theory that we could experiment with to test. When such an opinion is advanced, it is usually If it were possible to determine in advance with certainty how much love is inappropriate and how much love is excessive, then there would be a valid theory that we could experiment with to test it. When such an opinion is offered, it is usually the case that if it were possible to determine in advance with certainty how much love is inappropriate and how much love is excessive, then there would be a valid theory that we could experiment with to determine it to test. When such an opinion is put forward, it usually is Problems cannot be identified with certainty". Yes, but then you can't claim to know anything about it.

You will be shocked to hear that there are such examples in physics. We have symmetries that work more or less like this. We have approximate symmetry; and you compute a set of results, assuming it's perfect. There is no complete agreement when comparing it with the experiments. The symmetry you can expect is approximate. Therefore, when the fit is good, "Good!" you say. If the cohesion is too weak, you say, "This thing must be particularly susceptible to symmetry errors." Now you can laugh; but this is how we must move forward. When a subject is new, when these particles are new to us, the first step in any science is to "interrogate" it in order to predict the results. What applies to psychology also applies to the symmetry laws in physics; Don't laugh too much for him. In the beginning you have to be very careful. With a theory this vague, it's easy to fall into the deep side of the pool. It's hard to prove otherwise, and it takes skill and experience not to fall off the trampoline.

In this process, which consists of estimating, calculating and comparing the results with experimental data, we can get stuck where we are at some stages. When we can't generate thoughts, we're stuck in the guessing phase; It also happens during calculation. Yukawa for example[29] In 1934 he made an estimate of the intranuclear forces. However, because his math was so difficult, no one could calculate his results. Therefore, it was not possible to compare his claims with experimental results. These theories took a long time to discover all these particles that Yukawa didn't take into account, and things went wrong with what Yukawa thought.

Turns out it wasn't that easy. Another point where one can get stuck is the experimental aspect. For example, the quantum theory of gravity, if it makes any progress, makes it very slowly. Because none of the experiments you can do involve both quantum physics and gravity. The gravitational force is much weaker than the electric force.

Now that I'm a theoretical physicist and find this aspect of the problem more enjoyable, I want to focus on how predictions are made.

As I mentioned before, it doesn't matter where the guess comes from; It is important that it fits the experiment and is as precise as possible. They will say: "Then our job is very simple. Let's get a big calculator; Let be any disc repeatedly making predictions in it. Whenever the machine guesses a hypothesis about how nature works, it should immediately calculate its results and compare them with the list of experimental results at the other end." Guessing, then, is the business of idiots. But actually the reality is the opposite. Let me explain why.

The first problem is how to start. "I'll start with known principles," you will say. But the known principles are all mutually incompatible; therefore some should be excluded. We receive many letters emphasizing the need to leave gaps in our estimates. Because it is necessary to leave gaps to make room for new predictions. Someone tells us, "You know, you always mean that space is continuous. If you take a small enough size, there are enough points in between, small How do you know it's not a bunch of dots spaced apart?" Or they say, "How complicated and absurd are those quantum mechanical amplitudes you're talking about. How do you know they're right? Maybe they are she's not right." Such words are familiar to everyone who deals with this problem. It's no use pointing it out again. The problem is not just what cannot be right, but what can be safely put in its place. Suppose the correct statement about the continuity of space is that space consists of a series of points, the distances between them do not matter, and the points are arranged cubically. We can quickly prove that wrong. The problem is not just pointing out that something is wrong, but to replace it with something else; that's not that easy. Once you really put something specific in its place,

A second difficulty is that there are infinite possibilities of this simple kind. A situation like this: You are sitting and trying with great effort to open the safe; You worked a long time. Someone who doesn't know what you're doing except when you're trying to open a safe comes up to you and says, "Why don't you try the 10:20:30 combination?" says. You worked very hard, tried a lot, maybe you tried 10:20:30. Maybe you already know the middle number is 32, not 20... Maybe you know it's a five digit combination... Please don't send me letters telling me how to do this. I read them, I always read them to make sure what's proposed isn't something I hadn't thought of before. However, the answer takes a very long time; They usually belong to the Try 10:20:30 category. There are other theories that are very profound and mysterious.

As we have seen, nature's imagination is far superior to our own. It is not easy to make such a deep and mysterious prediction. It takes a lot of intelligence to guess; you can't do that blindly with a machine.

Now I want to talk to you about the art of discovering the laws of nature. This is an art. How it is done? One could argue that one way is to look back and see how others are doing. Let's look at the story.

We must start with Newton. He was faced with a situation where his knowledge was incomplete. He guessed the laws by putting his thoughts together, which came pretty close to experiment; there were no major differences between observations and experiments. This was the first method; but now it doesn't work so well.

The person who did something great after Newton was Maxwell, who discovered the laws of electricity and magnetism, what he did was this: he compiled all the laws of electricity that Faraday and other people before him had discovered, examined them and realized that they were mathematically inconsistent. To fix it, he had to add a term to the equation. And these are the idlers spinning in space.[30] and by designing a model composed of other types of gears. So he found a new law. But no one cared; because they didn't believe in idlers. Today we don't believe in these gears either; but the equations he found were correct. Even if the logic is wrong, the answer can be right.

The situation is quite different with the discovery of the theory of relativity. A pile of paradoxes had accumulated; known laws yielded conflicting results. A new way of thinking emerged from the discussion of possible symmetries in laws.

The situation was particularly difficult; for it was the first time that it was recognized that a law which might hold for a long time, like Newton's law, might be wrong. It was difficult to admit that the familiar notions of space and time that were intuitively perceived were also wrong.

Quantum mechanics was discovered in two independent ways (which is another lesson to remember). Here, too, a multitude of empirical contradictions and things that were absolutely inexplicable with the known had arisen. This was not because the information was incomplete, but because it was too complete. 'What you predicted would be so' was not the case, one of two different methods was Schrödinger's estimation of equations.[31], the other from Heisenberg, who suggested that we need to analyze the concept of measurability. These two different philosophical methods eventually led to the discovery of the same thing.

The recent discovery - although still not fully understood - of the weak fission laws I mentioned earlier involving the dissociation of a neutron into a proton, an electron and an antineutrino leads to a different situation. The problem with this is that the information is not complete, only the equations have been estimated. The particular difficulty is that all experiments give false results. If the result you calculated contradicts the experimental results, how can you predict where the error lies? It takes courage to say experiments are wrong. I will explain later where this courage comes from.

Today we have no more paradoxes; at least we think so. There is the problem of infinity that arises when we consider all the laws together. But people who sweep the

garbage under the rug are so smart that sometimes you think it's not a big paradox. And all those particles
The fact that we discovered it tells us that our knowledge is incomplete. If you think about the examples I've given, you'll no doubt realize that history in physics doesn't repeat itself. Let me explain it this way: Methods like "think about the laws of symmetry", "express information mathematically" or "guess equations" are now well known and used all the time. Why can't it be one of these when problems arise; because you must have tried them first. This time we have to find another way. If the multitude of difficulties and problems has led us to an impasse, it is because the methods we use are similar to those we have used before. The new project the new discovery is carried out in a completely different way. It seems that history doesn't help us much.

I want to tell you about Heisenberg's idea that one shouldn't talk about things that cannot be measured; because a lot of people talk about it without fully understanding it. We can interpret this idea as follows: your discoveries or inventions should be such that the results you calculate can be compared with the experiment. So if no one knows what moo or goo is, there is no point in calculating the result as "one moo is three goo". It is only important that the results are comparable with the experiment. It doesn't matter if Moss and Goo aren't included in the prediction. While expressing the prediction, you can use whatever words you want out of place. as long as the results are comparable. This point is not always well understood. Particles, orbit, etc. It is often lamented that concepts are transferred to the atomic world as soon as they are formed. This complaint is unfounded and there is nothing wrong with generalizations. beyond what we know and the ideas we have we always have to lie down; and we expand. Is it dangerous? Yes. Certainty missing? Yes. But it's the only way forward. It is necessary to make science useful even if it lacks certainty. Science is only helpful when it tells you about something untested; There's no point in just talking about things that happened. For example, in the law of gravitation, designed to understand the movements of the planets, if Newton had not simply said: "I know the planets now" and did not think that this could be compared to the earth's gravity on the moon, and then not leading others to believe that "gravity holds the galaxies together." it wouldn't be very helpful. We should try that. You could say: " that all conduct conforms to known laws; Experiments also give negative results. What we are looking for are hypotheses of this kind, precise and easily comparable with experiments. The fact is that nothing to the contrary has been observed in the behavior of galaxies. that all conduct conforms to known laws; Experiments also give negative results. What we are looking for are hypotheses of this kind, precise and easily comparable with experiments. The fact is that nothing to the contrary has been observed in the behavior of galaxies.

I can give you another, even more interesting and important example. The assumption that has perhaps made the greatest contribution to the development of biology is this: atoms can do everything that animals do, and that everything seen in the world of biology is the result of physical and chemical events, without "something extra". "Anything can happen when it comes to living beings," one might say. If you accept this, living beings You can't understand the time. It is very difficult to look at the wiggling of an octopus' tentacles and believe that this is nothing more than a movement of atoms in accordance with the laws of physics that we know. However, when examined in light of this hypothesis, fairly accurate predictions can be made as to how it works. In this way, great advances are made in the field of information gathering. So far, this hypothesis has not been proven wrong, nor have the tentacles been clipped.

Even if those not involved in the scientific world think otherwise, predictions are nothing against science. Years ago I had a conversation with an ordinary person about flying saucers. I should have known everything about flying saucers because I'm a "scientist"! "I don't think there are flying saucers," I said. The one before me said, "Is it impossible for flying saucers to exist? Can you prove it's impossible?" She asked. "No, I can't prove it's impossible; only the probability is very small," I said. But this is the scientific way. Being scientific just means to say what is probable and what is less probable; don't always try to prove the possible and the impossible.

How can we predict what to keep and what to discard? Despite what we know and our fine principles, we still do
We're in trouble: either we get infinities or we don't have enough definitions; there is a defect. Sometimes that means we have to discard some ideas. At least in the past, some strongly held ideas had to be discarded. The question is what to take and what to throw away. Throwing everything away would have gone a little too far. Then we don't have much to do. The principle of conservation of energy looks solid but nice; I don't want to throw it away. It takes a lot of skill to predict what to throw away and what to keep. In reality it might just be luck; but it still seems to require skill.

Probability amplitudes are very "strange things"; The first thing that comes to mind is that new ideas are distorted. However, the results of the quantum mechanics thesis that asserts that probability amplitudes exist are 100% correct for all of the strange particles on our list. So I don't think these concepts will prove wrong as we learn about the inner workings of the world. I believe this part is correct; But that's just a guess.

On the other hand, I believe that the theory that space is continuous is wrong; for we meet the infinities and other difficulties of which I have spoken; and the question arises as to what determines the size of all these particles. In my opinion it is wrong to carry the simple rules of geometry infinitely small into space. Here, too, I leave a gap and do not say what to put in its place. If I had, I would have ended this lecture with a new law.

Some people find contradictions in all principles; that there can only be a coherent world if we add up all the principles and

They used it to say that when we make very precise calculations, we not only identify principles, but that these principles are understood as the only ones that allow everything to remain consistent. This seems like an exaggeration; reminds me to grab the dog's tail and wag it. In my opinion one should admit that some things exist - electrons etc., if not all of the 50 or so particles. A few little things like - then the complexity that comes with all these principles may bring a certain result. I don't think everything can be explained by the consistency argument.

Another problem we have is what partial symmetries mean. These symmetries, such as neutrons and protons, are roughly the same but not electrically identical; Statements like or that the law of reflection symmetry is perfect except for a slight reaction are very annoying. Something is almost symmetrical, but not perfectly symmetrical. There are two schools of thought on this subject. The first suggests that the problem is simple, that things are indeed symmetrical, but that they're a bit "squinted" by a little confusion. A second school that has only one representative - that's me - also says no; itself can be complex, and only through this complexity does it become simple. The ancient Greeks thought that the orbits of the planets were circular. In reality they are ellipses. They're not perfectly symmetrical; however, they are very close to full circle. The question is; Why are they near the circle? Why are they almost symmetrical? For a long, complex reason like tidal friction. It may be that the nature of these things is inherently completely asymmetrical; but it seems almost symmetrical in the complexity of the facts, and Ellipses are like circles. This is another possibility; just a guess.

Suppose we have two theories, A and B, which appear to be completely different psychologically and have different ideas, but all their calculated results are identical and agree with the experiment. Although the two theories appear different at first, they lead to the same conclusions. This is easy to prove mathematically; It suffices to show that the logical content of A and B always gives the corresponding results. We suppose there are two such theories; How do we determine which is correct? There is no scientific way because both are equally compatible with the experiment. So these two theories are mathematically identical, although they stem from completely different ideas, and there is no scientific method

However, for psychological reasons, they can be far from equivalence in predicting new theories; because one may evoke different thoughts than the other. If you take an idea within a certain framework, you have an idea of what to change. For example, if there is a statement in Theory A that mentions something, you would say, "I'll change my mind over there." However, figuring out what to replace in B that matches it can be a very complicated task. In other words, a change that seems natural to one, even though they were identical before the change, is not natural to the other. From a psychological point of view, therefore, we must keep all theories in mind; a good theoretical physicist also knows six or seven different ways

He keeps them all in mind and hopes they will conjure up different ideas to implement.

This reminds me of something else: small changes in theory lead to big changes in philosophy and thinking around that theory; For example, Newton's ideas about space and time produced results that were very compatible with experiments. However, there was a slight difference in Mercury's orbit. In order to find the right movement in orbit, the core of the theory had to be fundamentally changed. This was because Newton's laws were very simple and precise, and gave certain precise results. They had to be completely changed to get a slightly different result. If you're trying to invent a new law, you can't add small flaws to perfection. Again, you need to find something perfect. This shows,

What are these philosophies? Indeed, these are clever methods to calculate results quickly. A philosophy, sometimes interpreted as perception of the law, is nothing more than a way of keeping an eye on the law in order to predict the consequences. Some people have made the following argument, which applies to situations like Maxwell's equations: "Ignore the philosophy; just guess the equations. The problem is to compute the answers in a way that is compatible with the experiments. There's no need for any philosophy, reasoning, or a word about the equation." If you're just guessing at the equation, it's good to get screwed up; It predicts the equation without bias, allowing you to make a better estimate. In contrast, philosophy can help with guessing. Making a decision is a difficult task.

For those who insist that the only thing that matters is the harmony between theory and experiment, I would like to quote a dialogue I had between a Mayan astronomer and his student. The Mayans for example, solar and lunar eclipses, the position of the moon in the sky, the position of Venus, etc. They were able to make fairly accurate predictions of matters and they did it all using arithmetic.

There was not the slightest dispute as to what the moon was; It wasn't even mentioned that he had returned. They just calculated things like when their eclipse would be, when it would be full moon.

Suppose a teenager goes to the astronomer and says, "I have an idea. These things can rotate and there can be balls of stone-like things there. Instead of calculating when they appear in the sky, we can calculate how they move in a very different way." The astronomer said to her, "Good. With what sensitivity can you calculate their eclipses?" says. He replied, "I haven't developed the calculations very well yet," to which the astronomer replied, "We can calculate lunar eclipses much more precisely than you can calculate with your model. You shouldn't dwell on these thoughts too much; because our mathematical model is obviously much better," he says. If someone comes up and suggests something, for example: "Let's imagine the world is like this," they say: "What is your conclusion for this and that problem?" Reaction tendencies are very strong. When he says, "I haven't developed the problem far enough," they say to him, "but we have, and we can get answers with a lot more certainty." As you can see, there is a problem of whether or not to think about the philosophy behind the ideas.

Another way of working, of course, is

to anticipate new principles. Einstein, in his theory of gravitation, all other principles In addition, he predicted the principle that corresponds to the principle that forces are always proportional to masses, he predicted the principle that in an accelerating car we cannot perceive anything other than being in a gravitational field, and by proving this principle with all the other principles combined he could find the right laws for gravity.

I have outlined some estimation methods for you. Now I would like to mention a few points about the results obtained. The first is what else can we do after we've done the work and have a mathematical formula that we can use to calculate the results. It's really amazing! To understand how an atom behaves in a given situation, we make rules by making marks on a piece of paper and giving them to a machine with complicated toggle switches; Finally, the machine tells us what the atom will do. If the way these buttons turn on and off is some kind of model of the atom, if we consider that atoms have buttons inside them too, then we have some understanding of what's going on. nothing to do with the basic I find it really amazing that you can predict what's going to happen using mathematical formulas made up of some rules. What happens in nature when buttons open and close on a computer is completely different.

One of the most important elements of this "guess – calculate results – compare experiment" process is knowing when we are right. It's possible to know we're right long before we've checked all the results. You can recognize the truth by its beauty and simplicity. After a guess and some small calculations, clearly showing that he's not wrong, it's easy to see that he's right.
If your guess is correct, it's obviously correct - at least if you're experienced - because what usually happens is; Much of what goes in goes out. Your guess is that the matter is very simple. If you don't immediately see that it's wrong and easier than before, you're right. Inexperienced people and the like also make simple guesses; but you soon realize that they are wrong. Therefore we do not take them into account. Some people, such as B. inexperienced students, however, make

predictions that are very complex and appear to be correct. I easily know they are wrong; because the truth is always easier than you think. What we need is imagination; but imagination in a terrible straitjacket. We need to find a whole new perspective on the world and this perspective should be compatible with all that is known. But the predictions must contradict somewhere; otherwise it would be uninteresting. However, it should be in harmony with nature, which contradicts it. When you find a point of view that agrees completely with everything observed but contradicts it elsewhere, you have made a great discovery. It is compatible when tested in relation to all theories tested, but gives different results in a different context; it is almost impossible to find even a theory whose results do not agree with nature; but not completely. Coming up with a new theory is extremely difficult and requires extraordinary imagination. but if you find a contradictory point of view elsewhere, you have made a great discovery. It is compatible when tested in relation to all theories tested, but gives different results in a different context; it is almost impossible to find even a theory whose results do not agree with nature; but not completely. Coming up with a new theory is extremely difficult and requires extraordinary imagination. but if you find a contradictory point of view elsewhere, you have made a great discovery. It is compatible when tested in relation to all theories tested, but gives different results in a different context; it is almost impossible to find even a theory whose results do not agree with nature; but not completely.

What does the future of this adventure look like? What will happen in the end? We keep guessing laws; How many more laws do we have to find? I do not know. Some of my colleagues say that this fundamental aspect of science will last forever; but I don't think this constant innovation will last for, say, a thousand years. This is the right way, that is, constantly finding new laws It can not go on like this. Still, it would be very tedious to have so many layers one below the other. I think one of two things could happen in the future. The first has all the laws, so there are enough laws, you can calculate the results, and the results all agree with the experiments. This means the end of the road. Or experiments become more difficult and expensive and you only find 99.9% of the laws; but there is always a phenomenon that contradicts something yet to be discovered and is very difficult to measure; As soon as you explain it, a new one comes out; things are getting slower and slower and less interesting. This is the other way to the end. However, I think it will end one way or another.

We are fortunate to live in a time when we can still make discoveries. This work is like the discovery of America; can only be done once. The age we live in is an age of discovering the laws of nature, and those days will never come again. It's extremely exciting, it's wonderful, but that excitement has to go. Of course, other interesting topics will appear in the future. Links between park-level phenomena will be of interest, eg phenomena in biology, etc.; or, if you want expeditions, expeditions to other planets. But the kind of things we're doing now won't exist.

For the future I foresee the following: when everything is finally known or it becomes too boring, the attention and active philosophy devoted to the topics discussed so far will gradually disappear. We will be surrounded by philosophers always standing outside and talking silly. Then we won't be able to prevent them by saying, "If you were right, we would have found all the rest of the laws." Because if all laws are in the middle, there will be suitable explanations for it. For example, it has always been explained why the earth is three-dimensional. Since there is only one earth, it is very difficult to understand whether these explanations are true or not. When all is known, it explains why these are the correct laws. However, these statements are made in a different framework, which we cannot criticize because this way of thinking prevents us from going further. Thoughts will degenerate in the same way that great explorers feel an area has degenerated with the arrival of tourists.

In our time, people are experiencing a great enthusiasm, an enthusiasm for making predictions about how nature will behave in an unprecedented situation. By using knowledge and experimentation in a given area, we can predict what will happen in an area that has never been explored before. This is a little different from the discoveries of explorers as we know them; where they have discovered they have enough clues to indicate what the undiscovered locations are. Our estimates, on the other hand, require a lot of thought. Often they are unlike anything we have seen before.

What does this allow in nature, that is, what allows us to predict what will happen in the whole from what happens in a part of it? This is an unscientific question. I don't know how to answer him; Therefore, I will give an unscientific answer. This characteristic, I believe, comes from the simplicity of nature and the beauty that it brings.

Printed in Great Britain
by Amazon